STUDENT UNIT G

NEW EDITIO

AQA A2 Biology U

Control in Cells and in O

Steve Potter and Martin Rowland

Philip Allan Updates, an imprint of Hodder Education, an Hachette UK company, Market Place, Deddington, Oxfordshire OX15 0SE

Orders
Bookpoint Ltd, 130 Milton Park, Abingdon, Oxfordshire OX14 4SB
tel: 01235 827827
fax: 01235 400401
e-mail: education@bookpoint.co.uk
Lines are open 9.00 a.m.–5.00 p.m., Monday to Saturday, with a 24-hour message answering service. You can also order through the Philip Allan Updates website: www.philipallan.co.uk

ISBN 978-1-4441-5295-1

First printed 2012
Impression number 5 4 3 2
Year 2017 2016 2015 2014 2013

Cover image: fusebulb/Fotolia

Printed in Dubai

Hachette UK's policy is to use papers that are natural, renewable and recyclable products and made from wood grown in sustainable forests. The logging and manufacturing processes are expected to conform to the environmental regulations of the country of origin.

P01966

Contents

Getting the most from this book

About this book

This guide will help you to prepare for **BIOL5**, the examination for **Unit 5: Control in Cells and in Organisms**, of the AQA A-level Biology specification. This examination will contain questions that are synoptic, so it is also important that you revise the principles listed at the end of all the other Biology units as well.

The **Content Guidance** section covers all the facts you need to know and concepts you need to understand for BIOL5. In each topic, the concepts are presented first. It is a good idea to make sure you understand these key ideas before you try to learn all the associated facts. The Content Guidance also includes examiner tips and knowledge checks to help you prepare for BIOL5. To help you prepare for questions in BIOL5 that are synoptic, some of the knowledge checks — indicated by Ⓢ — require you to use your understanding of topics from other units.

The **Questions and Answers** section shows you the sorts of question you can expect in the unit test. It would be impossible to give examples of every kind of question in one book, but these should give you a flavour of what to expect. Each question has been attempted by two students. Their answers, along with the examiner's comments, should help you to see what you need to do to score a good mark — and how you can easily *not* score a mark even though you might understand the biology.

What can I assume about this book?

You can assume that:
- the basic facts you need to know and understand are stated explicitly
- the major concepts you need to understand are explained clearly
- the questions at the end of the guide are similar in style to some of those that will appear in the BIOL5 unit test
- some of the questions test aspects of *How Science Works*
- the answers supplied are the answers of A2 students
- the standard of the marking is broadly equivalent to the standard that will be applied to your answers

So how should I use this book?

The guide lends itself to a number of uses throughout your course — it is not *just* a revision aid. You can use it:
- to check that your notes cover the material required by the specification
- to identify your strengths and weaknesses
- as a reference for homework and internal tests
- during your revision to prepare 'bite-sized' chunks of related material, rather than being faced with a file full of notes

You could use the Question and Answer section to:
- identify the terms used by examiners in questions and what they expect of you
- familiarise yourself with the style of questions you can expect
- identify the ways in which marks are lost as well as how they are gained

Develop *your* examination strategy

BIOL5 is the final examination of your A-level Biology course and its structure is different from the others you have taken. Consequently, you must develop a new examination strategy to prepare for BIOL5. But, be warned, developing your strategy is a highly personal and long-term process; you cannot afford to leave it to the last few weeks of your course.

Things you *must* do

- Clearly you must understand the topics covered in Unit 5 and be confident in applying your understanding to unfamiliar contexts. If not, you cannot expect to get a good grade.
- Understand the weighting of assessment objectives that examiners *must* use in BIOL5. They have designed BIOL5 with the approximate balance of marks shown in the table.

Assessment objective	Brief summary	Marks in BIOL5
AO1	Knowledge and understanding	27
AO2	Application of knowledge and understanding	57
AO3	How Science Works	16

- Understand where in BIOL5 different types of questions occur:
  - The final question will be a synoptic essay. It has a tariff of 25 marks — 16 for scientific content, 3 for breadth, 3 for relevance and 3 for quality of written communication. This essay carries many of the 27 marks for AO1 in BIOL5.
  - The penultimate question, worth 15 marks, will contain a high proportion of the AO3 marks and it too will be synoptic.
  - The remainder of the questions will test the topics in Unit 5 but will do so mainly in the context of AO2 skills; very few AO1 marks will be available here.
- Plan how you will attempt to answer the questions. Since you are now nearing the end of your course, you have, hopefully, developed all the skills that examiners test. You might, however, still have some that are better developed than others — skills that, in the pressure of an examination, you know will not let you down. You know the structure of the examination paper, so you can decide to attempt first those questions that test the skills in which you have the greatest confidence. If you like to write at length about what you have learnt, why not start with the essay? If your strongest skill is analysing and evaluating experimental data, why not start with the penultimate question?
- Use questions from past papers, from your textbook or from websites to maintain all the skills that examiners will test. Remember, the more you practise, the better your 'performance' will be.

Content Guidance

The Content Guidance section is a guide to the content of **Unit 5: Control in Cells and in Organisms**. It contains the following features.

Key concepts you must understand

Whereas you can learn facts, these are ideas or concepts that may form the basis of models that we use to explain aspects of biology. You can know the words that describe a concept, like negative feedback or transgenic organism, but you will not be able to use this information unless you really understand what is going on. Once you genuinely understand a concept, you will probably not have to keep re-learning it.

Key facts you must know and understand

These are exactly what you might think: a summary of all the basic knowledge that you must be able to recall and show that you understand. The knowledge has been broken down into a number of small facts that you must learn. This means that the list of 'Key facts' for some topics is quite long. This approach, however, makes quite clear *everything* you need to know about the topic.

Summary

This describes the skills you should be able to demonstrate after studying each related topic. These include the skills associated with the assessment objectives that examiners will ask you to demonstrate in the BIOL5 unit test.

Content Guidance

How organisms detect and respond to changes in their environment

The survival value of behaviour patterns

Key concepts you must understand

Organisms increase their chances of survival by responding to changes in their external or internal environment.

The processes shown in Figure 1 are common to the way that all organisms respond to their environment. There is always:

- a **stimulus** — a change in the organism's external or internal environment
- a **receptor** — a structure that detects the stimulus
- an **effector** — a structure, such as a muscle, that produces the response
- some kind of **linking system** or **coordinating system** — this is stimulated by the receptor and, in turn, stimulates the effector
- a **response** — the action that results from the stimulus

Knowledge check 1

Define the term *stimulus*.

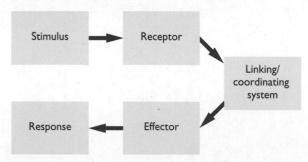

Figure 1 The processes involved in responding to a stimulus

Key facts you must know and understand

Responses of plants to external stimuli often involve growth. These responses are called **tropisms** and are directional responses to a directional stimulus. A growth response *towards* a stimulus is a positive tropism; a growth response *away* from a stimulus is a negative tropism. For example:

- plant shoots grow towards the most intense source of light (positive **phototropism**)
- plant shoots grow away from gravity (negative **gravitropism**)

Simple animals respond to external stimuli in one of two ways:

- A **taxis** (plural **taxes**), in which the animal moves along a gradient of intensity of a stimulus. Movement towards the stimulus is a positive taxis; movement away from the stimulus is a negative taxis. Maggots move away from light; male moths are attracted by, and move towards, pheromones (chemicals similar to hormones) released by female moths.
- A **kinesis** (plural **kineses**), in which a change in the intensity of the stimulus brings about a change in the rate of movement, *not* a change in the direction of movement. Woodlice move faster and change direction less often in dry conditions than in moist conditions. As a result, they are more likely to move out of dry conditions and remain in moist conditions, where they are less likely to dehydrate.

Taxes and kineses are examples of instinctive, non-intelligent behaviour.

Simple reflex actions in mammals are also instinctive, i.e. they are unlearned and the stimulus always produces the same response. There are two main types of reflex action:

- a **somatic reflex** is a response to an external stimulus — for example, the withdrawal reflex (page 28)
- an **autonomic reflex** is a response to an internal stimulus — for example, the reflexes controlling heart rate and breathing rate (page 11)

Variation in results

Investigating the behaviour of animals produces results with more variability than the results from an investigation into, say, enzyme activity. There are two main reasons for this:

- Only 20 or so woodlice are used, rather than billions of molecules of enzyme; one woodlouse that behaves differently has a bigger effect on the results than one enzyme molecule that behaves differently.
- There are genetic differences between woodlice and some of these influence behaviour.

You could investigate the preference of woodlice for a dark or a light environment using a simple choice chamber, such as the one in Figure 2.

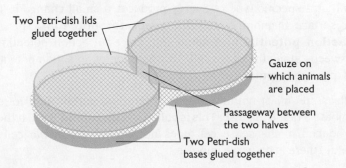

Two Petri-dish lids glued together

Gauze on which animals are placed

Passageway between the two halves

Two Petri-dish bases glued together

Figure 2 A simple choice chamber

You would place ten woodlice in each half of the choice chamber, one half of which is covered with black paper, and leave the apparatus for 10 minutes. At the end of this time, you would record the number of woodlice in each environment (dark and light). To help overcome the small sample size, you would repeat the experiment several times and calculate the mean number for each environment.

You could express your results as a bar chart, like the one in Figure 3. To give some idea of the variability of the results, you could calculate the standard error (SE) of each mean. Figure 3 shows error bars with values of 1.96 × SE above and below the mean values.

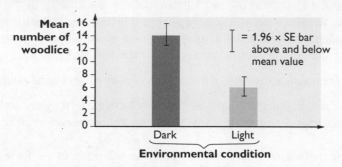

Figure 3 Bar chart of choice-chamber results

If the error bars for the two values overlap, you must conclude that the probability of the differences being due to chance is greater than 0.05 and they have no underlying cause.

If the error bars do not overlap, you must conclude that the probability of the differences being due to chance is less than 0.05 and there *is* an underlying cause for the differences.

Receptors

Key concepts you must understand

In humans, receptors are specialised sense cells that act as **energy transducers**. This means that a specific type of energy produces a small change in the voltage across their surface membrane, called a **generator potential**. This might then initiate an **action potential** in a nerve cell. Once an action potential has been initiated, it sweeps along the nerve cell. This passage of an action potential along a nerve cell is a **nerve impulse**.

Nerve impulses are 'all-or-nothing' events. There is a certain level of stimulation needed to initiate a nerve impulse. This is called the **threshold** level. If the generator potential produced in the sense cell does not stimulate a nerve cell above the threshold value, there is no nerve impulse.

Examiner tip
Notice the way in which the terms 'probability' and 'chance' are used here and use them in this way yourself. In section A of BIOL6, you would gain 2 marks for writing, appropriately: 'The probability that the differences in my results were due to chance is greater/less than 0.05, so I accept/reject my null hypothesis'.

Knowledge check 3
An action potential and a generator potential both involve a change in voltage (potential difference) across a cell surface membrane. How do they differ?

Detecting changes in the internal environment

Key concepts you must understand

Nearly all the reactions that take place in your body are controlled by enzymes. It is important that the enzymes work with maximum efficiency. They do this because you maintain them in an environment that is at, or very near to, their optimum temperature and pH. You also control other factors at appropriate levels.

Maintaining a constant internal environment is called **homeostasis** and involves **negative feedback** systems. You will find more details on both these concepts on pages 39–46.

To be able to maintain a constant internal environment, it is essential that you detect any changes. The receptors that do this are specific; they detect only changes in one particular factor.

Knowledge check 4

Suggest what makes receptors specific.

Key facts you must know and understand

Humans have receptors that detect changes in many factors, including:
- core body temperature
- plasma glucose concentration
- blood pressure
- the partial pressure of carbon dioxide in the plasma

You will learn about control of core body temperature and of plasma glucose concentrations on pages 40–44.

Control of heart beat

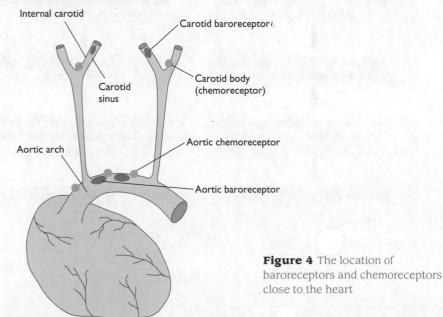

Internal carotid

Carotid baroreceptor

Carotid body (chemoreceptor)

Carotid sinus

Aortic arch

Aortic chemoreceptor

Aortic baroreceptor

S Knowledge check 5

Your blood pressure and partial pressure of CO_2 (pCO_2) both change during exercise. Explain what causes these changes.

Figure 4 The location of baroreceptors and chemoreceptors close to the heart

Changes in blood pressure and changes in the partial pressure of carbon dioxide in the plasma are important in the regulation of heart rate. **Baroreceptors** monitor changes in blood pressure; **chemoreceptors** monitor changes in the partial pressure of carbon dioxide in the plasma. The location of these sensors is shown in Figure 4.

The flow charts in Figure 5 show how changes in the partial pressure of carbon dioxide (pCO_2) and in arterial pressure influence the heart rate.

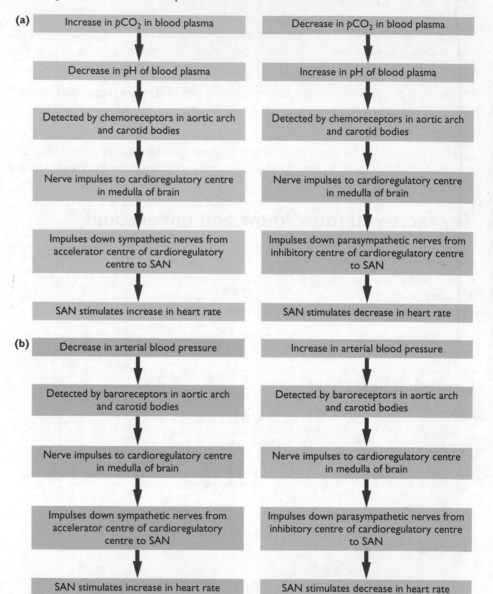

Figure 5 (a) The effect of a change in pCO_2 on heart rate (b) The effect of a change in arterial blood pressure on heart rate

S Knowledge check 6

You learnt about the SAN in Unit 1. Where, precisely, is the SAN?

Detecting changes in the external environment: pressure on the skin

Key concepts you must understand

The **Pacinian corpuscle** is one of several skin receptors that detect pressure.

Notice in Figure 6 the lamellae of a Pacinian corpuscle surrounding the end of a sensory nerve cell. Increased pressure applied to the Pacinian corpuscle distorts these lamellae and opens pressure-sensitive sodium ion channels in the membrane of the sensory nerve cell.

Since there is normally an excess of sodium ions (Na^+) outside, opening the sodium ion channels causes sodium ions to enter the nerve cell. Because sodium ions are positively charged, this movement of ions causes the change in voltage across the membrane — the **generator potential** — shown in Figure 7.

> **S** **Knowledge check 7**
>
> Name the process by which sodium ions enter the nerve cell. Use information from the passage to explain your answer.

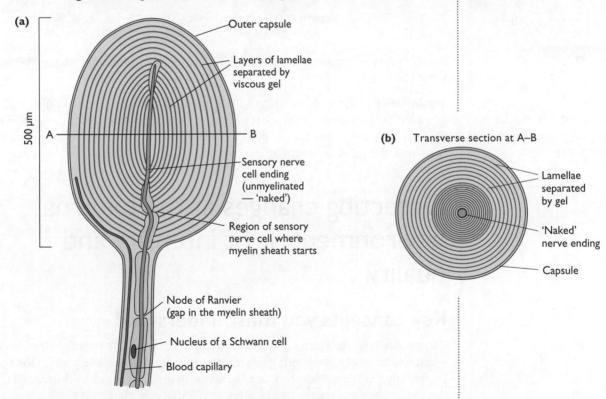

(a)
- Outer capsule
- Layers of lamellae separated by viscous gel
- 500 μm
- A — B
- Sensory nerve cell ending (unmyelinated — 'naked')
- Region of sensory nerve cell where myelin sheath starts
- Node of Ranvier (gap in the myelin sheath)
- Nucleus of a Schwann cell
- Blood capillary

(b) Transverse section at A–B
- Lamellae separated by gel
- 'Naked' nerve ending
- Capsule

Figure 6 (a) A Pacinian corpuscle seen in longitudinal section
(b) A Pacinian corpuscle seen in transverse section

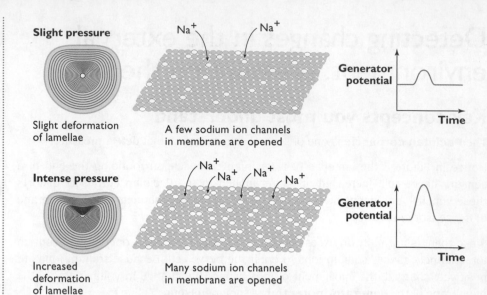

Slight pressure

Slight deformation of lamellae

A few sodium ion channels in membrane are opened

Generator potential

Time

Intense pressure

Increased deformation of lamellae

Many sodium ion channels in membrane are opened

Generator potential

Time

Figure 7 The effect of pressure on the generator potential produced by a Pacinian corpuscle

Greater pressure means that more sodium ion channels open and a bigger generator potential results.

The lamellar structure of the corpuscle is important in ensuring that only quite firm pressure stimulates the naked nerve ending. Light pressure deforms the lamellae slightly, but most of this pressure is then absorbed by the gel and not transmitted to the nerve ending at the centre of the corpuscle.

Detecting changes in the external environment: light intensity and quality

Key concepts you must understand

Receptor cells in the **retina** of the eye contain photosensitive pigments that degrade when struck by light. **Rod cells** contain **rhodopsin** and **cone cells** contain **iodopsin**. Degradation of the pigment produces a change in the voltage across the membrane of the rod or cone and results in a generator potential (see Figure 8).

Rods and cones differ in their:

- **acuity** — their resolving power
- **sensitivity** — the intensity of light that produces a sufficiently large generator potential to result in a nerve impulse (action potential). Rhodopsin is much more sensitive to light than is iodopsin, i.e. it degrades in dimmer light than does iodopsin.

Examiner tip
You can refer to the voltage across a membrane or to the potential difference across a membrane; an examiner will accept either.

S Knowledge check 8
You learnt about resolving power in Unit 1. What does it mean here?

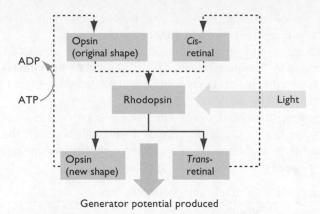

Figure 8 How rhodopsin produces a generator potential when struck by light

Both these differences are relevant to the way in which the rods and cones synapse with (are linked to) cells called **bipolar cells**. This is illustrated in Figure 9.

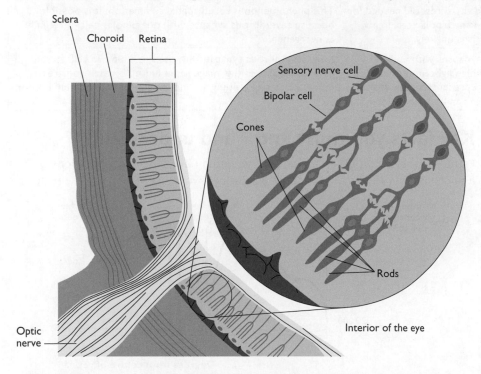

Figure 9 The way in which rods and cones synapse with bipolar cells in the retina

Each cone cell synapses with one bipolar cell, which then initiates a nerve impulse in one sensory nerve cell. There are two consequences of this:

- Cones are relatively insensitive to dim light. Consequently, in dim light, the generator potential will not reach the threshold needed to stimulate the bipolar cell.
- The area of the visual field detected by the single cone is interpreted by the brain as a single point (think of a pixel) in the image it produces. This gives the cones good **visual acuity**.

Several rod cells synapse with the same bipolar cell. This is called **retinal convergence** and there are two consequences of this:

- The generator potentials from several rod cells combine or 'add up' to reach the threshold needed to stimulate the bipolar cell. This is called **spatial summation** — it makes it more likely that the threshold will be exceeded in dim light and that impulses will pass to the brain.
- Light striking several rod cells results in a single impulse along just one sensory nerve cell to the brain. The brain cannot distinguish between the parts of the image produced by the different rods and interprets them as one point. Therefore, rods have poor acuity.

Knowledge check 9

What is (a) retinal convergence; (b) spatial summation?

The table below provides a simple comparison of cones and rods.

Property	Cones	Rods
Sensitivity	Low since iodopsin degraded only by bright light. Light energy transduced by a single cone must produce a generator potential large enough to exceed the threshold needed for a nerve impulse in the bipolar cell. In low light intensities this is unlikely.	High since rhodopsin degraded by dim light. In low light intensities, generator potentials from several rods can combine and so the threshold is more likely to be exceeded and a nerve impulse initiated in the bipolar cell. This phenomenon is called spatial summation. It is possible because several rods synapse with one bipolar cell — retinal convergence.
Acuity	High: each cone synapses with a single bipolar cell, so in high light intensities each cone stimulated represents a separate part of the image which can be seen in detail.	Low: several rods synapse with the same bipolar cell, so the individual parts of the image represented by each rod are merged into one — they are indistinguishable and detail is poor.

Key facts you must know and understand

Figure 10 shows that different parts of the retina have different proportions of rods and cones. This affects the kind of image formed when light falls on those regions.

Figure 10 The distribution of rods and cones in the retina

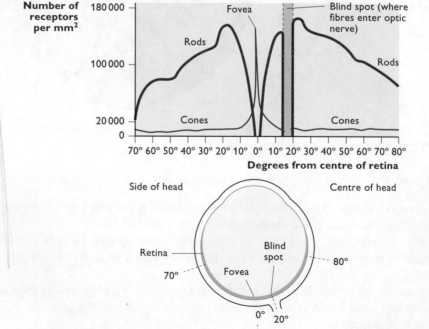

Knowledge check 10

Use Figure 10 to describe the distribution of rods in the retina.

The greatest concentration of cones is found at the **fovea** in the centre of the retina. When you look straight at an object, light from the object is focused onto your fovea. This enables you to see the object in great detail (provided the light intensity is sufficiently high).

There are three different types of cone. Each type is sensitive to the different wavelengths of light that are broadly equivalent to the three primary colours — red, blue and green. Rods are equally sensitive to all wavelengths of light, so colour is perceived only when light falls on the cones.

Knowledge check 11

At night, you see a faint star better if you look slightly to one side of it than if you look directly at it. Explain why.

Summary

After studying this topic, you should be able to:
- interpret given data to explain how organisms increase their survival chances by responding to changes in their external and internal environments
- state that receptors are highly specific in the stimulus to which they respond
- distinguish between a kinesis, a taxis and a tropism
- describe the role of baroreceptors and chemoreceptors in controlling the heart rate
- explain how pressure causes a generator potential in a Pacinian corpuscle
- use the terms retinal convergence and spatial summation in explaining differences in the sensitivity and visual acuity of different regions of the retina

How organisms coordinate their responses to stimuli

Organisation of the human nervous system

Key facts you must know and understand

The nervous system is composed of billions of cells. Most of these cells are involved in transmitting impulses and are called **neurones**. They are organised into larger structures including nerves, the spinal cord and the brain.

The brain and the spinal cord form the **central nervous system** (**CNS**); the nerves form the **peripheral nervous system** (**PNS**). Figure 11 shows this organisation.

The nervous system is divided functionally into:
- the **somatic nervous system** (SNS), which controls responses to external stimuli
- the **autonomic nervous system** (ANS), which controls responses to internal stimuli

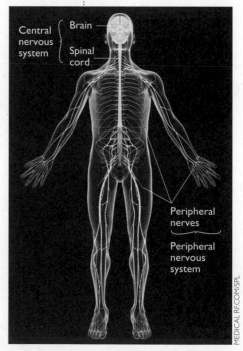

Figure 11 The structural organisation of the nervous system

The ANS is further subdivided into:

- the **sensory branch**, which transmits sensory nerve impulses into the central nervous system
- the **sympathetic branch**, which transmits impulses from the central nervous system to the organs, generally preparing the body for 'fight or flight', e.g. by increasing cardiac output and pulmonary ventilation
- the **parasympathetic branch**, which acts *antagonistically* to the sympathetic branch and prepares the body for 'rest and repair', e.g. by decreasing cardiac output and pulmonary ventilation

The functional organisation of the nervous system is shown in Figure 12.

Knowledge check 12

The sympathetic and parasympathetic branches of the autonomic nervous system work *antagonistically*. Explain what this means.

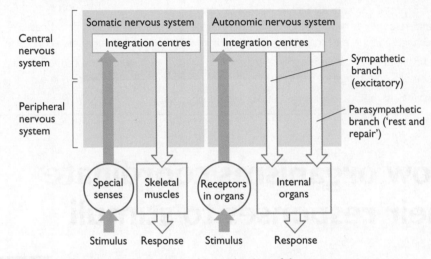

Figure 12 The functional organisation of the nervous system

The nervous system contains:

- **sensory neurones** — these carry nerve impulses from receptors into the CNS
- **motor neurones** — these carry impulses from the CNS to effectors such as muscles
- **relay neurones** — these carry impulses from a sensory neurone to a motor neurone, often in **reflex arcs**, within the CNS

Examiner tip

Do not confuse the terms *nerve cell* and *nerve*. A nerve cell, or neurone, is a single cell; a nerve is a bundle containing the axons of a large number of neurones.

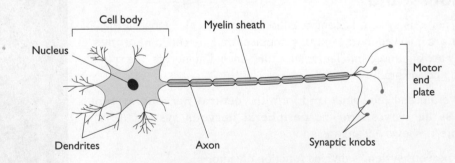

Figure 13 A myelinated motor neurone

The motor neurone shown in Figure 13 is well adapted for transmitting nerve impulses. It has:

- many small extensions called **dendrites**, which are stimulated by other neurones within the CNS

- a long process called an **axon**, which transmits impulses from the CNS to a muscle or gland
- a **myelin sheath** around its axon. This is formed by **Schwann cells** that wrap themselves around the axon as the motor neurone develops. You can see the effect of this in the upper part of Figure 14. The surface membrane of each Schwann cell contains myelin — a lipid that effectively insulates the axon against leakage of ions and makes nerve impulses travel faster along the axon.
- synaptic knobs at the ends of the axon, which stimulate the target effector

Examiner tip
An examiner will expect you to write that axons transmit impulses. Avoid writing that they transmit 'messages' or 'information'.

The resting potential

Key concepts you must understand

When not conducting an impulse, an axon has a membrane potential called its **resting potential**. In this state, the inside of the membrane is more negatively charged than the outside by about 70 millivolts (70 thousandths of a volt). This is usually written as –70 mV. Because of this difference in charge between the inside and outside of the membrane, we say the membrane is **polarised**.

Key facts you must know and understand

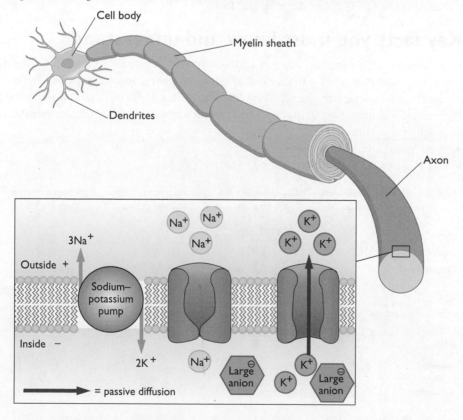

Figure 14 The resting potential

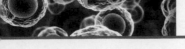

S Knowledge check 13

In which direction along their concentration gradient are sodium ions transported by sodium–potassium pumps?

Figure 14 shows how the resting potential is maintained by:

- large anions, for example negatively charged protein molecules inside the axon
- facilitated diffusion of sodium ions (Na⁺) and potassium ions (K⁺) across the membrane through ion channels. Sodium ions diffuse in more slowly than potassium ions diffuse out so the inside of the membrane becomes more negative.
- active transport of sodium and potassium ions across the membrane by carrier proteins, called **sodium–potassium pumps**. More sodium ions are pumped out than potassium ions are pumped in (again, the inside of the membrane loses positively charged ions faster than it gains them).

Depolarising the membrane

Key concepts you must understand

An action potential is initiated when either a receptor or another neurone secretes a chemical (**neurotransmitter**) that crosses an excitatory **synapse** (small gap) between the secreting cell and the neurone in which the action potential will be initiated.

The neurotransmitter binds to receptors on the membrane of the neurone and causes its membrane potential to change momentarily from a negative value to a positive value. This change is called **depolarisation**.

Key facts you must know and understand

Depolarisation happens because, in addition to the structures shown in Figure 14, the surface membrane of a neurone also contains **gated ion channels**. These ion channels can be open or closed. They are firmly closed during the maintenance of the resting potential. Figure 15 shows that, when a tiny part of the surface membrane of a neurone is stimulated:

- some gated sodium ion channels open
- all the gated potassium ion channels remain closed

Examiner tip

Candidates often write that 'sodium ions rush into a neurone'. This is poor use of terminology; it is better to write that sodium ions rapidly enter the neurone down a steep diffusion gradient.

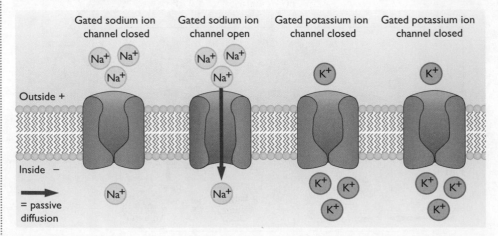

Figure 15 When an action potential is initiated, some gated sodium ion channels open, but all the gated potassium ion channels remain closed

Opening the gated sodium ion channels allows sodium ions to diffuse into the axon. This makes the membrane potential less negative. If the membrane potential reaches a **threshold** of about –55 mV, yet more gated sodium ion channels open. Because they only open at this threshold voltage, they are called **voltage-sensitive sodium ion channels**.

When all the voltage-sensitive sodium ion channels open, sodium ions rapidly diffuse into the cytoplasm. All the extra (positively charged) sodium ions inside the membrane raise its potential to about +40 mV (with respect to the outside of the membrane). You can see the effect of this in the graph in Figure 16. The 'spike' in the curve is called an **action potential**.

Once the membrane potential reaches a particular value (+40 mV in Figure 16), all the gated sodium ion channels close. No more sodium ions can enter the axon and the membrane potential cannot become any more positive.

A consequence of the threshold potential described above is that if depolarisation only causes the membrane potential to rise to (say) –60 mV, the voltage-sensitive sodium ion channels do not open and there is no action potential. An action potential is an **all-or-nothing** event.

Examiner tip
Resting potentials and action potentials have different values in different animals. Don't worry if you see a question in which a graph has values that are different from the ones used here. Similarly, you will not be penalised if you write that an action potential has a value of, say, +30 mV.

Repolarising the membrane

Key concepts you must understand

The axon membrane cannot initiate another action potential until it has returned to its 'resting' condition, i.e. it has **repolarised**.

Key facts you must know and understand

As soon as the action potential (+40 mV) is reached and the gated sodium ion channels close, the gated potassium ion channels open. Consequently, positively charged potassium ions diffuse rapidly out of the cytoplasm. This restores the resting potential of –70 mV (with respect to the outside).

In practice, too many potassium ions diffuse out of the cell and reduce the membrane potential to about –80 mV. This 'undershoot' is called **hyperpolarisation**. It is quickly restored to the resting potential of –70 mV.

The events of depolarisation and repolarisation are shown in Figure 16. It is important to realise that this graph shows changes that occur *over a short period of time* at *one place* in the axon.

There is a period of time following the initiation of one action potential during which another action potential cannot be generated. This is called the **refractory period** and includes:
- the <u>absolute</u> refractory period (the periods of depolarisation and repolarisation)
- the <u>relative</u> refractory period (the period of hyperpolarisation and re-balancing of sodium and potassium ion concentrations)

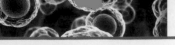

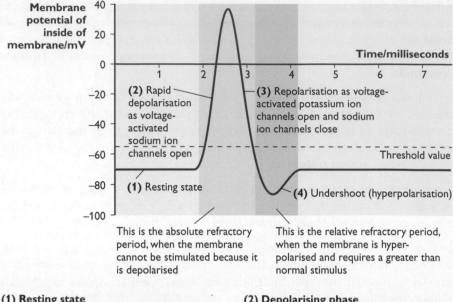

This is the absolute refractory period, when the membrane cannot be stimulated because it is depolarised

This is the relative refractory period, when the membrane is hyper-polarised and requires a greater than normal stimulus

(1) Resting state

Outside axon

Na⁺ Gated potassium channel

+ + + + + +   + +   + + + +

Gated sodium channel

Inside axon   K⁺

(2) Depolarising phase

Outside axon

Na⁺

− − − − − −   − −   − − − −

+ + + + + +   + +   + + + +

Inside axon   K⁺

(3) Repolarising phase

Outside axon

Na⁺

+ + + + + +   + +   + + + +

Inside axon   K⁺

(4) Undershoot

Outside axon

Na⁺

+ + + + + + +   + +   + + + +

Inside axon   K⁺

Figure 16 The depolarisation and repolarisation cycle

The refractory period limits the frequency with which impulses can be transmitted along a neurone. Explain how.

Propagating action potentials along an axon

Key concepts you must understand

An impulse travels along an axon because an action potential at one point of the surface membrane causes an action potential further along the membrane. We say that the action potential is propagated.

You have seen that at the point of the axon where an action potential occurs, the inside of the axon becomes positively charged. This causes negatively charged ions ahead of the action potential to move to this positively charged area, i.e. it causes a local electric current in the cytoplasm. This, in turn, causes an action potential in the membrane ahead.

Propagation is faster in myelinated neurones than in non-myelinated neurones.

Key facts you must know and understand

Propagation along non-myelinated neurones

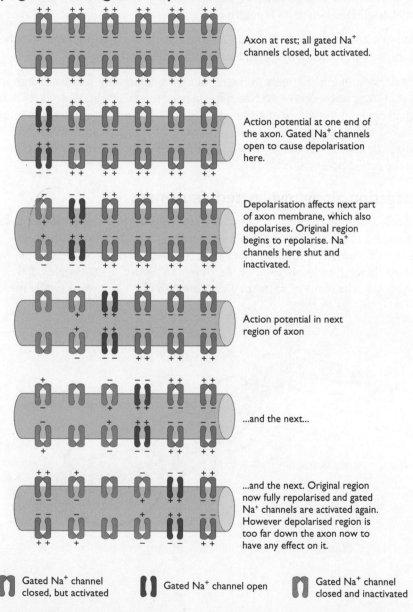

Axon at rest; all gated Na$^+$ channels closed, but activated.

Action potential at one end of the axon. Gated Na$^+$ channels open to cause depolarisation here.

Depolarisation affects next part of axon membrane, which also depolarises. Original region begins to repolarise. Na$^+$ channels here shut and inactivated.

Action potential in next region of axon

...and the next...

...and the next. Original region now fully repolarised and gated Na$^+$ channels are activated again. However depolarised region is too far down the axon now to have any effect on it.

Gated Na$^+$ channel closed, but activated

Gated Na$^+$ channel open

Gated Na$^+$ channel closed and inactivated

Figure 17 Propagation of an action potential along a non-myelinated neurone

Gated ion channels can exist in three states:
- open — appropriate ions can pass through
- closed but activated — ions cannot pass, but the gate will open when stimulated
- closed and inactivated — ions cannot pass and the gate will not open when stimulated (think of it as being 'locked')

The top of Figure 17 shows that, in the 'resting' state, all the gated ion channels are closed, but activated.

S Knowledge check 15

The speed with which an impulse is transmitted is affected by temperature. Explain why.

When an action potential is generated at one end of an axon, the depolarisation affects the axon membrane just ahead of it. Figure 17 shows that this new region now begins to depolarise and generate an action potential as the region where the action potential originated begins to repolarise.

The action potential in the new region now affects the axon membrane just ahead of it and this begins to depolarise...and so on down the axon.

S Knowledge check 16

A non-myelinated neurone with a large diameter conducts impulses faster than a non-myelinated neurone with a smaller diameter. Suggest why.

The gated sodium ion channels in the repolarising region are inactivated as well as being shut. They only become activated again once the membrane is fully repolarised. As you can see in the lowest diagram in Figure 17, by this time the action potential has moved some way down the axon; it cannot affect the area where the action potential originated. This explains why action potentials cannot move backwards to where the action potential originated.

Propagation along myelinated neurones

Look back to Figure 17. Now imagine Schwann cells around the axon. Because they insulate the axon, the Schwann cells ensure that depolarisation can only occur at the nodes of Ranvier — the gaps between Schwann cells. As a result, the action potentials 'jump' from node to node, as shown in Figure 18. This gives this method of transmission its name of **saltatory conduction** (the Latin word *saltare* means 'to jump').

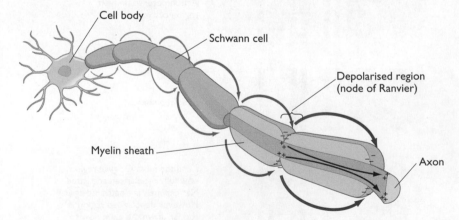

Figure 18 Saltatory conduction in a myelinated neurone

Knowledge check 17

Which type of neurone — myelinated or non-myelinated — will conduct impulses faster? Explain your answer.

Structure of the synapse

Key facts you must know and understand

A **synapse** is the structure that allows impulses to be passed from:
- a receptor to a neurone
- one neurone to another neurone
- a neurone to an effector (e.g. the neuromuscular junction in Figure 35)

It is not simply the gap between two cells, which is the **synaptic cleft**.

Where impulses are transmitted between two neurones:
- the neurone that transmits impulses is the **pre-synaptic neurone**
- the neurone that receives impulses is the **post-synaptic neurone**

Transmission across the synaptic cleft occurs when substances called **neurotransmitters** are secreted from vesicles in the pre-synaptic neurone into the synaptic cleft.

Synapses may be named according to the specific neurotransmitter that is secreted.
- At a **cholinergic synapse**, **acetylcholine** is the neurotransmitter.
- At an **adrenergic synapse**, **noradrenaline** is the neurotransmitter.

Figure 19 shows the structure of a cholinergic synapse

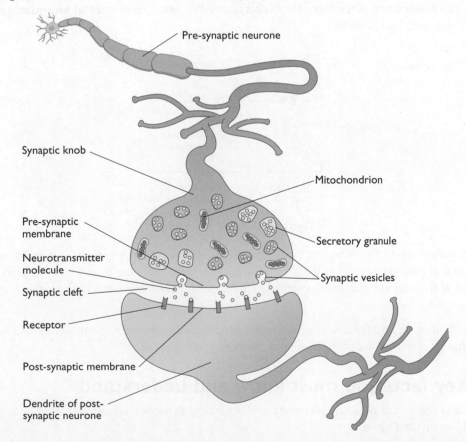

- Pre-synaptic neurone
- Synaptic knob
- Mitochondrion
- Pre-synaptic membrane
- Secretory granule
- Neurotransmitter molecule
- Synaptic vesicles
- Synaptic cleft
- Receptor
- Post-synaptic membrane
- Dendrite of post-synaptic neurone

Knowledge check 18

Which type of neurone passes an impulse to an effector?

S Knowledge check 19

A synaptic knob contains a large number of mitochondria. Explain the advantage of this large number of mitochondria.

Figure 19 Structure of a cholinergic synapse

Transmission across the synapse

Key concepts you must understand

Synapses can be excitatory or inhibitory.

Different neurotransmitters are secreted at excitatory synapses and inhibitory synapses.

- At an **excitatory synapse**, once the neurotransmitter (e.g. acetylcholine) has been secreted, it binds to receptors on the post-synaptic membrane and *decreases* the membrane potential (makes it *less* negative). If the threshold of about −55 mV is reached, an action potential will be initiated in the post-synaptic neurone (see page 20).

- At an **inhibitory synapse**, once the neurotransmitter (e.g. GABA — gamma amino butyric acid) has been secreted, it binds to receptors on the post-synaptic membrane and *increases* the membrane potential (makes it *more* negative). This makes it *less* likely that the threshold will be reached and *less* likely that an action potential will be initiated in the post-synaptic neurone.

Usually, several neurones (not just two) synapse together. As Figure 20 shows, some of these synapses may be excitatory and others inhibitory. The effect on the post-synaptic neurone depends on the combined effect of the neurones that release neurotransmitter at the same time. This is another example of **spatial summation** that you met on page 16.

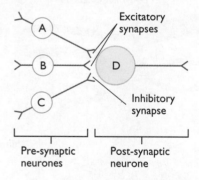

Figure 20 Spatial summation

Another type of summation is **temporal summation**. In temporal summation, several impulses arrive at the synapse from the same neurone in quick succession. Each depolarises the post-synaptic membrane a little more until the threshold is reached.

In both types of synapse, the neurotransmitter is rapidly hydrolysed. This ensures that its effect on the post-synaptic membrane is short-lived.

Key facts you must know and understand

The events that occur at excitatory and inhibitory synapses are summarised in the flowchart in Figure 21.

S Knowledge check 20

Some neurones respond to acetylcholine, others to GABA and yet others to noradrenaline. Explain this specificity.

Knowledge check 21

Look at Figure 20. Will an action potential be initiated in neurone D as a result of release of neurotransmitter by: (a) neurone A alone; (b) neurone A + neurone B; (c) neurone A + neurone C? Explain each of your answers.

S Knowledge check 22

In most synapses, the products of hydrolysis of the neurotransmitter are reabsorbed by the pre-synaptic membrane. Explain the advantage of this reabsorption.

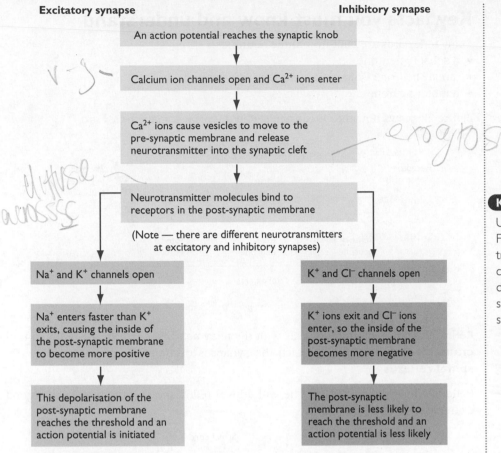

Figure 21 Differences between an excitatory and an inhibitory synapse

Knowledge check 23

Use information in Figure 21 to explain why transmission at synapses can only occur in one direction (from pre-synaptic neurone to post-synaptic neurone).

The organisation of neurones in reflex arcs

Key concepts you must understand

Reflex arcs are pathways that allow simple reflex actions to be carried out.

A reflex arc always:
- receives stimuli from a specific receptor
- transfers impulses to a specific effector

Knowledge check 24

Explain what is meant by a *simple reflex*.

There are, broadly speaking, two main kinds of reflex action:
- **Somatic reflexes** involve our special senses (eyes, ears, pressure receptors etc.) and produce a response by a muscle — for example, the 'withdrawal-from-heat' reflex.
- **Autonomic reflexes** involve sensors in internal organs and produce responses in internal organs — for example, the control of heart rate and breathing rate.

Key facts you must know and understand

Many reflex arcs are built from three neurones:

- a sensory neurone
- a relay neurone (sometimes called an interneurone)
- a motor neurone

Figure 22 shows how they are organised in a typical somatic reflex arc.

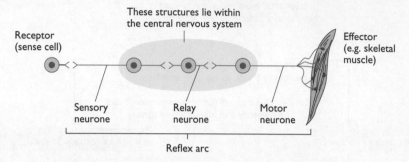

Figure 22 The structures in a somatic reflex arc

Reflex arcs in which the synapses with the relay neurone occur in the brain control **cranial reflexes**; those in which the synapses occur in the spinal cord control **spinal reflexes**.

Reflex actions are protective. The withdrawal reflex shown in Figure 23 is a good example.

Figure 23 The withdrawal reflex

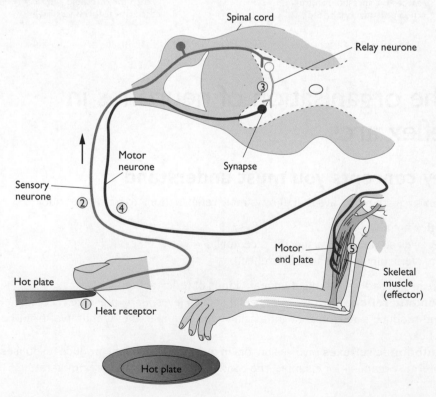

The numbers in Figure 23 refer to the following sequence of events:

1 Heat from the hot plate causes a generator potential in a heat receptor in the finger.

2 The generator potential initiates an action potential in a sensory neurone, which is transmitted along the neurone.

3 At the synapse with a relay neurone, the sensory neurone releases neurotransmitter that initiates an action potential in the relay neurone.

4 This is repeated at the synapse between the relay neurone and a motor neurone and an action potential is transmitted along the motor neurone.

5 At the motor end plate, the action potential stimulates the contraction of the skeletal muscle, causing the automatic withdrawal of the hand from the hot plate.

Autonomic reflexes often occur in antagonistic pairs; one controlled by the sympathetic branch of the ANS and the other by the parasympathetic branch. The control of heart rate, which you learnt about in Figure 5, involves:

- reflex arcs in the sympathetic branch of the ANS that increase heart rate
- reflex arcs in the parasympathetic branch of the ANS that decrease heart rate

Knowledge check 25

The reflex arc shown in Figure 23 does not involve your brain, yet you are aware of the heat. Suggest how impulses reach your brain.

Control by hormones

Key concepts you must understand

As we have seen, nerve cells stimulate their target cells by secreting a neurotransmitter directly onto them across a synaptic cleft.

In contrast, hormones are substances that are secreted into the blood by **endocrine glands** and stimulate cells in other, distant parts of the body. Figure 24 shows that, although they reach all cells via the blood, hormones stimulate only cells with **receptor proteins** that are complementary to the shape of the hormone.

Knowledge check 26

An examination question contains a graph showing strength of response plotted against time. It has two curves, one representing nervous control and the other representing hormonal control. Give *two* ways in which you could tell which curve is which.

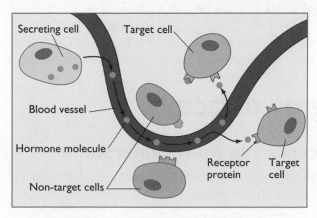

Figure 24 How hormones target specific cells

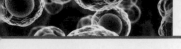

Knowledge check 27

Explain why steroids can pass freely through cell membranes.

Examiner tip

If asked to explain how hormonal control differs from nervous control, don't forget that the examiner wants reasons. Hormonal action is slower *because* the hormone takes time to reach its target via the blood; its effects are widespread *because* the hormone gets to all cells via the blood; its effects are longer lasting *because* the hormone is not quickly hydrolysed.

Examiner tip

If asked to make a comparison, make sure you compare like with like. You *would* gain credit for writing that prostaglandins have a local effect whereas hormones have a widespread effect. You would *not* gain credit for writing that prostaglandins have a local effect whereas hormones have a slow effect.

Examiner tip

You must make clear that vasodilation (and vasoconstriction) occurs only arterioles.

The chemical nature of a hormone affects the way it stimulates its target cells.

- Steroids (e.g. sex hormones) are small, lipid-soluble molecules that can pass freely through plasma membranes and bind with receptors in the cytoplasm. These receptor–hormone complexes move into the nucleus, binding with and activating specific genes.
- Non-steroids (e.g. glucagon) bind to receptors on the surface of the cell. This activates a molecule in the membrane called a 'G-protein', which activates specific enzymes that produce specific metabolic effects.

You will find specific examples of the ways in which the two types of hormone work on pages 45 and 42–43. You will also find how hormones interact in controlling the oestrous cycle on page 45.

Local chemical mediators: prostaglandins and histamine

Key concepts you must understand

Prostaglandins and histamine are **chemical mediators**. Like hormones, they are released by the cells that produce them and affect target cells. Unlike hormones, they act locally — only affecting cells in their immediate vicinity.

As you learnt in Unit 1, prostaglandins and histamine are involved in the inflammatory response to injury.

Prostaglandins are released by damaged cells throughout the body. They stimulate:
- arterioles to dilate (vasodilation) so that more blood flows to an infected or damaged area
- migration of phagocytic white blood cells to the area
- blood clotting

Histamine is released by mast cells in a damaged or infected part of the body. It stimulates:
- capillary walls to become more 'leaky', allowing tissue fluid, complement proteins and phagocytic white blood cells to leave capillaries more easily

Plant growth factors

Key concepts you must understand

In flowering plants, growth is controlled by specific plant growth factors that diffuse from dividing cells to other tissues.

The discovery of the action of auxins: how scientists take forward the research of other scientists

Early research into the action of auxins was carried out using coleoptiles — the specialised leaf that covers the emerging shoot of the seedling of a cereal.

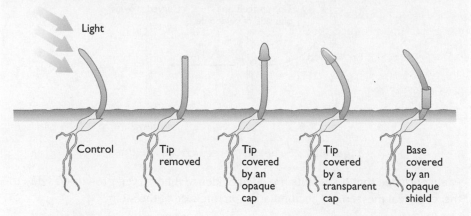

Figure 25 The results of research by Charles and Francis Darwin (1880)

Knowledge check 28

Explain how the results in Figure 25 support the Darwins' conclusion that light is detected by the tip of a coleoptile.

From the results of their experiment, shown in Figure 25, the Darwins concluded that:

- detection of the stimulus occurs in the tip of the coleoptile
- a region some distance behind the tip brings about the response
- there was communication between the two, probably chemical in nature

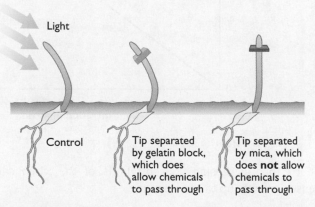

Figure 26 The results of research by Boysen-Jensen (1913)

Knowledge check 29

Explain how the results in Figure 26 support Boysen-Jensen's conclusion.

Boysen-Jensen concluded that the communication between detection and response was chemical in nature.

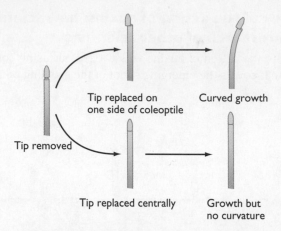

Knowledge check 30

Explain how the results in Figure 27 support Paal's conclusion.

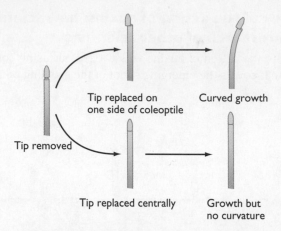

Figure 27 The results of research by Paal (1918)

Paal concluded that his results provided evidence that curved growth was due to the 'chemical messenger' accumulating on one side of the stem.

In 1926, Went investigated the effects of agar blocks containing different concentrations of the 'chemical messenger' on the growth of coleoptiles with their tips removed.

Knowledge check 31

Which investigation provides the most reliable evidence that a chemical messenger is involved in the control of growth in coleoptiles? Explain why.

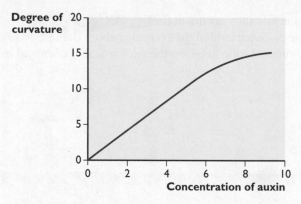

Figure 28 The results of research by Went (1926)

Examiner tip

An examiner could not expect you to recall any of these experimental results — they are not in the specification. An examiner would, however, expect you to be able to interpret and evaluate these results — a valid test of AO3.

Went concluded that increasing concentrations of the 'chemical messenger' caused increased growth, up to a maximum.

Since these experiments, the 'chemical messenger' has been identified as a group of plant growth factors called **auxins**, the major one of which is indoleacetic acid (IAA). Indoleacetic acid has been isolated and purified and shown to have exactly the same effects as the 'chemical messenger' first predicted by the Darwins.

Biologists believe that IAA acts on genes controlling growth, turning them on and stimulating both cell division and cell elongation (Figure 29).

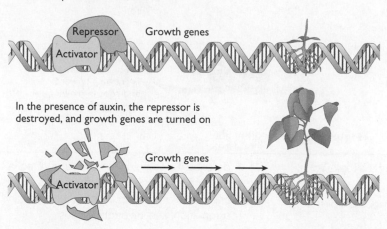

In the absence of auxin, plant growth genes are in a repressed, or off, state

Repressor   Growth genes

Activator

In the presence of auxin, the repressor is destroyed, and growth genes are turned on

Activator   Growth genes

Figure 29 Auxins, such as IAA, act by inactivating repressor substances that block the activation of growth genes in plants

The growth towards unilateral light is the result of IAA being redistributed to the shaded side of the shoot (Figure 30).

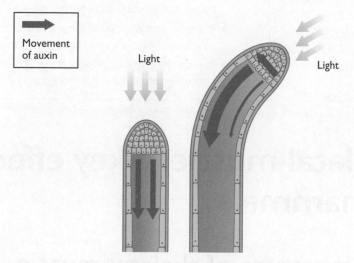

Movement of auxin

Light   Light

Figure 30 The phototropic response of a shoot. IAA moves away from the lit side of a shoot and accumulates in the shaded side. Cells in the shaded side elongate more as a result, causing a curvature towards the light.

The **gravitropic response** shown in Figure 31 is also controlled by IAA. Again, elongation of cells in a shoot is stimulated by IAA. Elongation of cells in a root is, however, inhibited by high concentration of IAA.

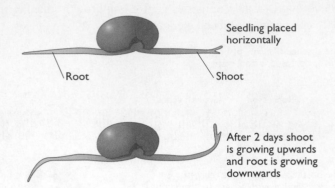

Seedling placed horizontally

Root

Shoot

After 2 days shoot is growing upwards and root is growing downwards

Figure 31 Investigating how a root and shoot respond to gravity

Summary

When you have completed studying this topic, you should be able to:

- show understanding of the organisation of the human nervous system
- describe the structure, and name the components, of a myelinated motor neurone
- use your understanding of membrane permeability to explain resting potential, deoplarisation, action potential, repolarisation, refractory period and the the all-or-nothing principle
- explain how a nerve impulse is transmitted along a neurone and how this transmission is affected by myelination, axon diameter and temperature
- describe the structure of a synapse
- describe the sequence of events involved in impulse transmission across a cholinergic synapse and use this to explain unidirectionality, temporal and spatial summation, and inhibitory synapses
- interpret unfamiliar information relating to nerve impulse transmission, including the effect of drugs on transmission across synapses
- describe a simple reflex arc involving three neurones and explain the importance of simple reflexes in avoiding damage to the body
- compare and contrast the features of coordination by nerves, hormones and chemical mediators
- explain the effect of indoleacetic acid (IAA) in controlling tropisms in flowering plants

Skeletal muscle: a key effector in mammals

The structure of skeletal muscle

Key concepts you must understand

Each end of a skeletal muscle is attached by inelastic tendons to bones. When the muscle contracts it pulls one of these bones into a different position.

A muscle contains a large number of muscle cells. Most are in the middle of the muscle, which is why it bulges. The structure of a single muscle cell — or fibre — is well adapted for its function. It is packed with protein filaments, which cause the cell to contract, and mitochondria, which produce the ATP needed during contraction.

Key facts you must know and understand

A muscle cell may be 30–300 mm long; for this reason it is called a **muscle fibre**. Figure 32 shows that a single muscle is made of bundles of muscle fibres, surrounded by connective tissue.

Each muscle fibre is surrounded by a surface membrane, called the **sarcolemma**. Its cytoplasm (**sarcoplasm**) contains:
- several, scattered nuclei
- a large number of mitochondria that produce the ATP used during contraction
- a network of tubules, the **sarcoplasmic reticulum**, that store the calcium ions (Ca^{2+}) that start muscle contraction

Most of the cytoplasm, however, is packed with much thinner fibres, called **fibrils**. These, in turn, contain **filaments** of two types of protein, thin **actin** filaments and thicker **myosin** filaments.

S Knowledge check 34

A muscle, such as the biceps, is classified as an organ. Explain why.

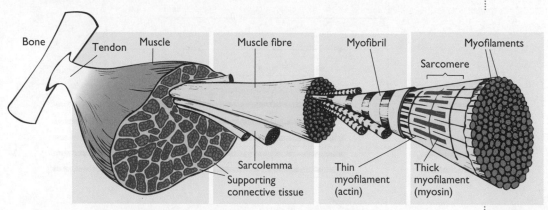

Figure 32 The structure of skeletal muscle

How skeletal muscle moves bones: the sliding filament theory

Key concepts you must understand

A skeletal muscle can only exert a force when it *contracts*; it can, therefore, only *pull* a bone. Reversing the movement of the bone involves the contraction of a second muscle. This explains the mechanical advantage of skeletal muscles working in **antagonistic pairs** — for example, the biceps and triceps muscles that flex (bend) and extend (straighten) the arm.

The generally accepted theory of muscle contraction is called the **sliding filament theory**. According to this theory, filaments of actin and myosin in the cytoplasm of a muscle cell slide over one another, shortening the cell. Increased strength of muscle contraction results from an increased number of cells contracting.

Examiner tip

If you place the components of a muscle in alphabetical order — fibre, fibril and filament — they are also in order of decreasing size. Remembering this might help you to use the correct terminology when sitting BIOL5.

Key facts you must know and understand

Shortening of sarcomeres

Look at the top half of Figure 33; it shows the working unit of a myofibril — the **sarcomere**.

At the edges of each sarcomere are two **Z lines**, to which thin filaments of actin are attached. These thin actin filaments produce a light band on either side of the Z line, called the **I band**. In the middle of each sarcomere is a dark band — the **A band** — caused by the thicker myosin filaments. Now look at the same sarcomere in the lower half of Figure 33. Can you see that the A band has stayed the same width but the I bands have almost disappeared? This has happened because the actin filaments have been pulled into the A band; the overlap between actin and myosin filaments is almost total.

Knowledge check 35

The H zone in Figure 33 appears lighter than the rest of the A band but darker than the I band. Explain why.

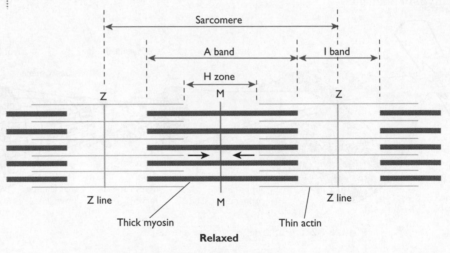

Figure 33 Shortening of a sarcomere according to the sliding filament theory

An outline of the mechanism

- The ends of the myosin filaments are bulbous and are called **myosin heads**.
- These heads bind with **binding sites** all along the actin filaments.
- As they bind, the heads tilt and move the actin filaments.
- The myosin heads are then released and return to their original positions.
- They bind to other sites further along the actin filament and repeat the tilting and moving.

Knowledge check 36

During muscle contraction shown in Figure 33, (a) have any of the labelled filaments changed their length, (b) have any of the labelled regions changed their length? If so, which?

The stages in muscle contraction

When a muscle is relaxed:
- molecules of **tropomyosin** block the binding sites on the actin filaments, preventing the myosin heads from binding
- each myosin head is in a 'resting position' and is bound to a molecule of ATP

When a muscle contracts, the following events, shown in Figure 34, occur.
- Calcium ions diffuse into the muscle fibres and cause the tropomyosin molecules to move and expose the myosin binding sites on the actin molecules.
- At the same time, the ATP molecules that are bound to the myosin heads are hydrolysed, to form ADP and inorganic phosphate (which is released). The energy released is transferred to the myosin heads.
- The myosin heads bind to the exposed binding sites on the actin molecules (sometimes referred to as **cross-bridge formation**), releasing the ADP.
- The energy previously transferred to the myosin heads now moves the heads and with them the actin filaments.
- Another ATP molecule attaches to each myosin head, which now detaches from the actin.
- Provided nervous stimulation continues to cause calcium ions to be released, the cycle continues. Otherwise, the muscle returns to the relaxed state.

Examiner tip

Quite a lot of detail is involved in explaining how muscle contraction takes place. Remember that, outside the essay, there are few AO1 marks in BIOL5 so make sure you understand what is going on so that you can apply your understanding to novel contexts.

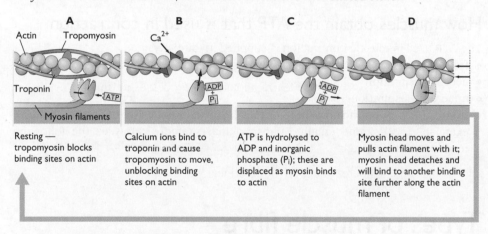

Figure 34 How the sliding filament theory explains muscle contraction

The neuromuscular junction

You can see that contraction of a muscle fibre starts with diffusion of calcium ions from its sarcoplasmic reticulum into its sarcoplasm. To explain what causes this, we have to look how motor neurones stimulate muscles.

Motor neurones synapse with muscle fibres at **neuromuscular** junctions, like the one shown in Figure 35. The following events stimulate muscle fibres to contract:
- Nerve impulses arrive at the neuromuscular junction.
- Calcium ions enter the synaptic knob, causing the release of acetylcholine into the synaptic cleft (all the neuromuscular junctions in skeletal muscle are cholinergic).
- Molecules of acetylcholine diffuse across the synaptic cleft and bind to specific receptors on the sarcolemma. This causes the sarcolemma to depolarise.

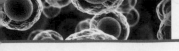
- Depolarisation of the sarcolemma causes calcium ions to be released from the sarcoplasmic reticulum into the fibrils. This influx of calcium ions starts the contraction process outlined above.

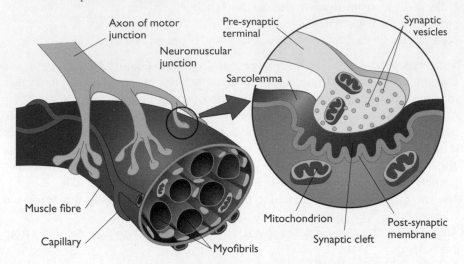

Figure 35 A neuromuscular junction

How muscles obtain the ATP that is used in contraction

The ATP used in muscle contraction is supplied in four ways:
- from 'stored' ATP — this lasts for about 3 seconds of intense exercise
- aerobic respiration ⎫
- anaerobic respiration ⎬ You learnt about these processes in Unit 4
- from stored **phosphocreatine** — this lasts for about 10 seconds of intense exercise. The enzyme creatine phosphokinase rapidly catalyses the following reaction in muscle cells:

ADP + phosphocreatine → ATP + creatine

Types of muscle fibre

Key facts you must know and understand

There are two types of muscle fibre.

Slow-twitch fibres are adapted for releasing energy aerobically over a sustained period. They have:
- many mitochondria in each fibre
- high concentrations of the enzymes that regulate the Krebs cycle (most reduced NAD and reduced FAD molecules — which release electrons into the electron transfer chains of the inner mitochondrial membrane — are produced in the Krebs cycle)
- a more extensive capillary network than fast-twitch fibres (allowing faster delivery of oxygen and removal of carbon dioxide)

Knowledge check 37

Calcium ions are involved in muscle contraction in *two* ways. Identify them.

Knowledge check 38

List in order of use the sources of ATP in the muscles of: (a) an athlete who sprints 100m in 12 seconds; (b) an athlete who runs a marathon in 4 hours. Explain your answers.

Knowledge check 39

Different people have different proportions of slow- and fast-twitch fibres in their muscles. Which type would you expect to be more common in the muscles of: (a) an athlete who sprints 100m in 12 seconds; (b) an athlete who runs a marathon in 4 hours? Explain your answers.

- a high concentration of **myoglobin**, a protein that stores oxygen
- a relatively slow contraction rate with less force than fast-twitch fibres
- a high resistance to fatigue

Fast-twitch fibres are adapted for releasing energy anaerobically over a short period of time. Fast-twitch fibres:

- have fewer mitochondria than slow-twitch fibres
- have high concentrations of the enzymes that control glycolysis (and low concentrations of the Krebs cycle enzymes)
- have a higher concentration of ATPase than slow-twitch fibres so that a lot of ATP can be hydrolysed quickly
- have a higher contraction rate, with more force, than slow-twitch fibres
- have low concentrations of myoglobin
- have a lower resistance to fatigue than slow-twitch fibres

Knowledge check 40

Different muscles within your own body have different proportions of slow- and fast-twitch muscle fibres. Which type would you expect to be more common in: (a) the muscles that close your eyelids; (b) the muscles in your back?

Summary

After completing this topic, you should be able to:

- describe the gross and microscopic structure of skeletal muscle
- use your knowledge of its ultrastructure to interpret the appearance of a myofibril in both a contracted and a relaxed state
- explain the roles of actin, ATP, calcium ions, myosin and tropomyosin during the contraction of a muscle fibril

- compare the structure, location and general properties of fast-twitch and slow-twitch muscle fibres
- interpret information relating to fast-twitch and slow-twitch muscle fibres, and to the sources of ATP, used during different forms of exercise
- describe a neuromuscular junction and explain how it stimulates muscle contraction

---

Homeostasis and feedback systems

Key concepts you must understand

Homeostasis involves physiological control systems by which a mammal maintains its internal environment within restricted limits. It ensures that the tissue fluid, which surrounds the cells of a mammal's body, has a near constant:

- temperature
- pH
- water potential

Maintaining a constant internal environment allows cellular enzymes to work in their optimum conditions and, therefore, carry out metabolic processes with maximum efficiency.

These factors are maintained within their limits by **negative feedback** systems. Figure 36 shows that such systems detect when the value of a particular factor changes outside its 'preset range' and activate mechanisms that bring it back within that range.

S Knowledge check 41

What is the importance to a mammal of maintaining a constant water potential in its tissue fluid?

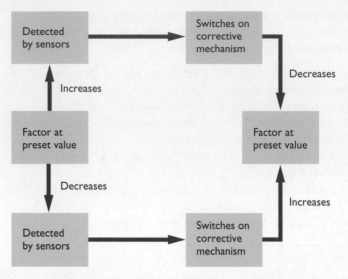

Figure 36 The principle of negative feedback

Some systems in the body operate using **positive feedback**. In such systems, a change is detected and the change is then amplified (made larger). The system only 'reverts to normal' when the stimulus is removed.

Key facts you must know and understand

Regulating core body temperature

Animals can be placed into two groups on the basis of how they regulate their core body temperatures:

- **Endotherms** use physiological processes and behavioural methods to regulate their core temperature.
- **Ectotherms** use only behavioural methods to regulate their core temperature.

Core body temperature will remain constant if heat gains balance heat losses.

Humans gain heat by:

- conduction — we touch something hotter than our own skin (or consume hot food)
- convection — hot circulating air warms our own skin
- radiation — objects hotter than us radiate heat
- respiration — much of the energy released in respiration is lost as heat; very active organs such as the liver and active skeletal muscle generate large amounts of heat

Humans lose heat by:

- conduction — we touch something cooler than our skin (or consume cold food)
- convection — we heat colder air around us
- radiation — we radiate heat from our skin
- evaporation — when water in sweat turns to a vapour on our skin at least part of the heat to make this happen comes from the skin

Core body temperature is regulated by a **thermoregulatory centre** in the **hypothalamus**. It is divided into two antagonistic centres:

Examiner tip

As with many biological concepts, the division of animals into ectotherms and endotherms is not absolute. Don't be confused if an examination question contains information that clearly shows a reptile having some ability to control its temperature by physiological means.

Examiner tip

You will not gain marks in an examination for writing that sweating cools the body. It is the *evaporation of sweat* that removes heat from the skin. You must say so.

- The **heat gain centre** activates measures that conserve heat and increase heat generation.
- The **heat loss centre** activates measures that increase heat loss and decrease heat generation.

The hypothalamus receives impulses about body temperature from two sources.
- Sensors in the skin detect *changes* in the temperature of the skin.
- Sensors in the hypothalamus itself detect *changes* in the temperature of the blood flowing through it, which reflect changes in the 'core' body temperature.

Figure 37 summarises the role of the hypothalamus in controlling body temperature.

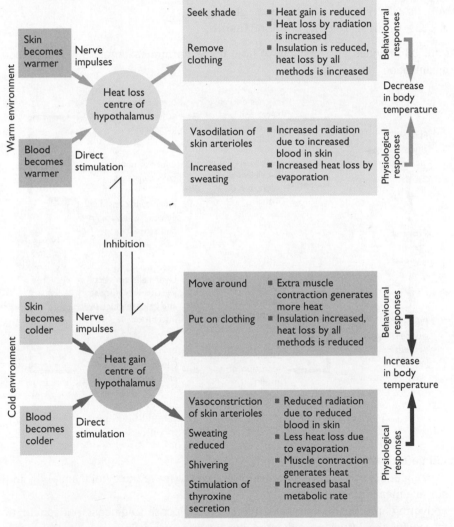

Basal metabolic rate is the rate of energy expenditure when a person is awake but resting, has not eaten for 12 hours and is comfortably warm.

Figure 37 Mechanisms involved in regulating core body temperature in humans

The hypothalamus 'orders' the regulatory processes through the **autonomic nervous system** (ANS). The two divisions once again act antagonistically.

The **sympathetic division** carries impulses from the heat gain centre to:
- increase metabolic rate and heat generation
- activate processes that reduce heat loss

The **parasympathetic division** carries impulses from the heat loss centre to:
- decrease metabolic rate and heat generation
- activate processes that increase heat loss

Controlling plasma glucose concentration

Small groups of cell in the pancreas, called **islets of Langerhans**, contain two different types of endocrine cell:
- α-cells, which secrete the hormone **glucagon**
- β-cells, which secrete the hormone **insulin**

Figure 38 provides an overview of how these hormones regulate plasma glucose concentration.

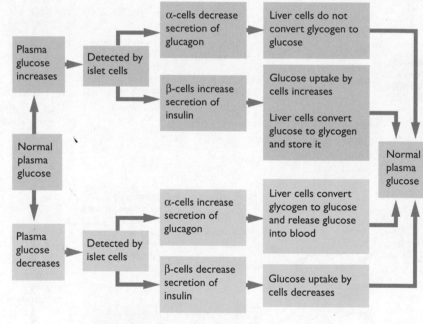

Figure 38 The role of insulin and glucagon in the regulation of plasma glucose concentration

Insulin affects glucose metabolism by:
- causing additional carrier proteins for glucose to leave the cytoplasm and join the surface membranes of liver cells
- activating enzymes in liver cells that convert glucose to glucose phosphate, thus increasing the diffusion gradient of glucose into liver cells
- activating enzymes that convert glucose phosphate into glycogen (**glycogenesis**)

Glucagon affects glucose metabolism by:
- activating enzymes in the liver that hydrolyse stored glycogen into glucose
- activating enzymes in the liver that convert other substances to glucose (**gluconeogenesis**)

Examiner tip
It might help you to recall in an examination which cells of the pancreas secrete which hormone if you remember that 'α' is the Greek letter 'a' and that, of the two hormones, only glucagon contains a letter 'a'.

S Knowledge check 42
Give *two* advantages of maintaining a constant blood glucose concentration.

Examiner tip
We have a number of words here with similar spellings, including glucagon, glucose and glycogen. Normally, an examiner will not be too harsh if your spelling is wobbly but here your spelling must be accurate to gain credit.

Although it does not have the same regulatory effect, adrenaline — the fright, fight or flight hormone — also stimulates the breakdown of glycogen to glucose and the release of glucose into the bloodstream.

Glucagon exerts its effect by a process called the **second messenger model**, shown in Figure 39.

- Molecules of glucagon bind to specific receptors that span the surface membranes of liver cells.
- This activates an enzyme on the inside of the membrane (the G-protein shown in Figure 39).
- The activated enzyme removes two of the three phosphate groups from ATP, producing cyclic adenosine monophosphate (**cAMP**).
- cAMP activates other enzymes that catalyse the breakdown of glycogen to glucose.

Adrenaline works in a similar way but binds to different receptor proteins and activates a different enzyme.

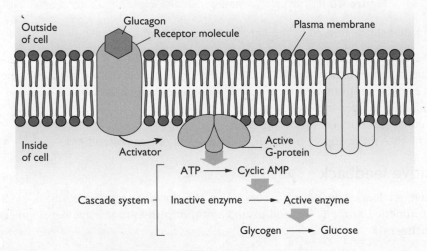

Figure 39 The second messenger model

Type 1 and type 2 diabetes

Diabetes is a condition in which the homeostatic control of plasma glucose concentration fails.

- **Type 1 diabetes** is caused by an inability to produce insulin. This results from the immune system destroying large numbers of the body's own β-cells early in life.
- **Type 2 diabetes** is caused by the plasma membranes of liver cells losing their sensitivity to insulin.

Sustained hyperglycaemia (high plasma glucose concentrations) results in two key changes in the body.

- Not all the glucose filtered from the blood by the kidneys is reabsorbed.
- Tissues become dehydrated as they lose water by osmosis to the blood (because of the decreased water potential of the plasma).

These changes are the cause of several of the symptoms of diabetes, as shown in Figure 40.

Knowledge check 43

Type 1 diabetes is often described as 'early-onset' diabetes and type 2 diabetes as 'late-onset' diabetes. Suggest why.

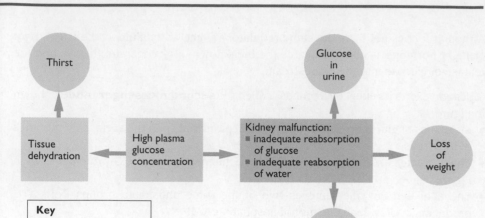

Figure 40 How hyperglycaemia causes the symptoms of diabetes

Type 1 diabetes is treated by regular injections of insulin.

Type 2 diabetes is treated by:
- a diet planned to avoid 'peaks' and 'troughs' in the concentration of plasma glucose
- an exercise programme to ensure that glucose in the plasma is used
- oral anti-diabetes medication to
  - reduce gluconeogenesis in the liver
  - enhance the production of insulin in the islets of Langerhans

Positive feedback

Positive feedback sometimes occurs when the normal negative feedback systems fail to function properly. A good example of this is the progressive development of **hypothermia**.

Usually, when body temperature falls, the body responds by increasing its metabolic rate. This results in more energy being released. However, if this is not enough to offset the heat losses, core body temperature continues to fall. Beyond a certain point, further reduction in body temperature *reduces* the metabolic rate, so *less* heat is produced. This means that the core body temperature now reduces more quickly — so the metabolic rate reduces more quickly, and so on.

Positive feedback and negative feedback systems control mammalian oestrous cycles

Not all positive feedback is harmful, though. Both positive and negative feedback systems safely control the oestrous cycle in female mammals. This is the cycle that controls:
- the maturation and release (ovulation) of oocytes (potential egg cells)
- changes in the uterus lining that enable implantation and growth of zygotes
- changes in the sexual behaviour of the female, which comes into oestrus (or 'heat')

The length of each oestrous cycle is different in different mammals. In mice, the cycle is five days long whereas red deer come into oestrus only once each year.

Examiner tip
You are *not* required to know about the changes that occur in the ovary or uterus during one oestrous cycle. You can only be tested on your understanding of the positive and negative feedback control of the hormone secretion.

The oestrous cycle in women is approximately one month long, hence the name menstrual cycle.

The events in this cycle are under the control of the four hormones shown in Figure 41 — two gonadotrophic hormones and two steroid sex hormones.

Gonadotrophic hormones are secreted by the anterior pituitary gland and they target cells in the ovaries.

- **Follicle stimulating hormone** (**FSH**) stimulates the development of one or more follicles (the structures in which oocytes develop) and stimulates them to secrete oestrogen.
- **Luteinising hormone** (**LH**) stimulates ovulation (the release of an oocyte) as well as stimulating cells in the follicle to secrete progesterone.

Steroid sex hormones are secreted by cells in the ovary. They target the uterine lining as well as stimulating and/or inhibiting the secretion of gonadotrophic hormones:

- **Oestrogen**
  - low concentrations of oestrogen inhibit the secretion of FSH
  - higher concentrations of oestrogen stimulate the secretion of FSH and of LH
- **Progesterone** maintains the uterine lining and inhibits the secretion of both FSH and LH.

> **Examiner tip**
> Note the noun is *oestrus* but the adjective is *oestrous*. You would not be penalised for using the wrong spelling.

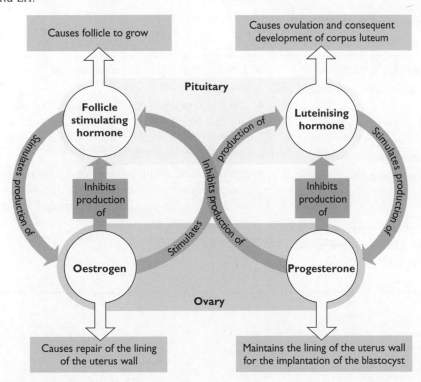

Figure 41 The interactions between the hormones controlling the oestrous cycle

Figure 42 represents these changes in hormone concentration as a graph; in this case showing a human menstrual cycle. It also shows physical changes in an ovary, though this is not in the specification.

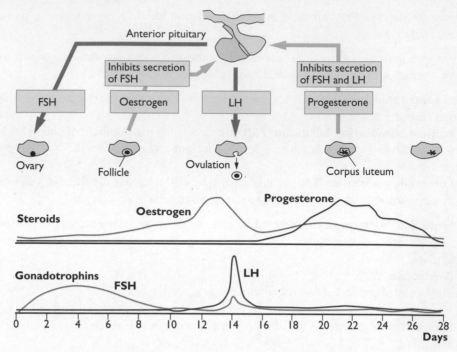

Look at Figure 42.
(a) Explain the shape of the curve for FSH between days 1 and 13. (b) Explain the peak in the curve for LH between days 13 and 15.

Figure 42 The changes in the concentrations of the hormones controlling the human menstrual cycle

Summary

After studying this topic, you should be able to:

- explain the concepts of homeostasis and negative feedback
- explain the importance of maintaining a constant core temperature, a constant blood pH and a constant blood glucose concentration
- contrast the mechanisms of temperature control in an ectothermic reptile and an endothermic mammal
- explain how mammals control their core body temperature and their blood glucose concentration, including the relevant nervous and hormonal control mechanisms

- compare the origin and control of type 1 diabetes and type 2 diabetes
- use the concepts of negative feedback and positive feedback to explain how FSH, LH, oestrogen and progesterone control the mammalian oestrous cycle
- interpret data, including those in diagrams and graphs, relating to the control of core body temperature, blood glucose concentration and the oestrous cycle

The genetic code, protein synthesis and gene mutation

A brief revision of the structure of DNA

The structure of DNA is shown in Figure 43.

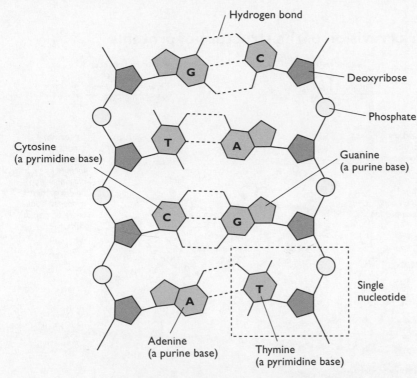

Note: adenine (A) always bonds with thymine (T)
guanine (G) always bonds with cytosine (C)

Figure 43 The structure of DNA

The DNA molecule is made from two strands arranged into a **double helix**:

- Each strand is a polynucleotide consisting of four types of **nucleotide**.
- The two strands are held together by hydrogen bonds.
- The two strands are **anti-parallel**.
- Only one strand of DNA holds coding information; this is the **coding (sense) strand**; the other strand is the **non-coding (antisense) strand**.
- Each nucleotide contains the pentose **deoxyribose**, a **phosphate group** and an **organic (nitrogen-containing) base**.
- Each polynucleotide strand is held together by the 'sugar–phosphate **backbone**' in which ester bonds form between the phosphate group of one nucleotide and the deoxyribose sugar of the next nucleotide.
- The four organic bases are **adenine**, **thymine**, **cytosine** and **guanine**.

- Specific base pairing occurs between the strands:
  - adenine is always paired with thymine
  - cytosine is always paired with guanine
- In addition to the coding regions (**exons**) within a gene, there are non-coding sections of DNA called **introns**, which separate the exons.
- There are also base sequences between genes that do not code for amino acids and which repeat themselves; these include **minisatellites** and **microsatellites**.

A brief revision of the structure of proteins

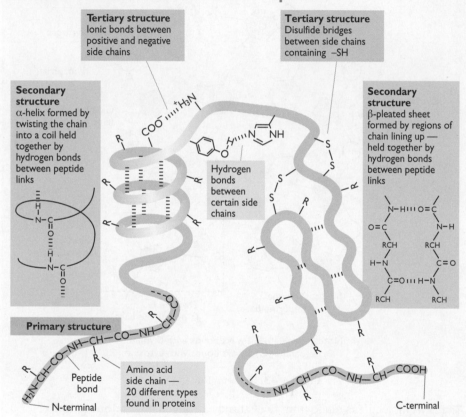

Figure 44 The primary, secondary and tertiary structures of a protein

Protein molecules are **polymers** of amino acids. The amino acids are joined together by **peptide bonds** to form a **polypeptide chain**. Each protein molecule has three levels of organisation:

- **primary structure**
- **secondary structure** (either an α-**helix** or β-**pleated sheet**)
- **tertiary structure**

These levels of organisation are described in the table below and shown in Figure 44.

Level	Bonds holding structure in place	Description	Notes
Primary	Peptide bonds	Sequence of amino acids in the poly-peptide chain	Determined by sequence of triplets of bases in DNA
Secondary	Hydrogen bonds	α-helix or β-pleated sheet	Formed by folding poly-peptide chain; both types can exist in the same protein molecule
Tertiary	Ionic, hydrogen and disulfide bridges	Globular or fibrous structure	Gives molecule unique shape and specific function

The genetic code and protein synthesis

Key concepts you must understand

A **gene** is a region in the coding strand of a DNA molecule that codes for a particular protein. The code for the protein is determined by the sequence of organic bases within the gene.

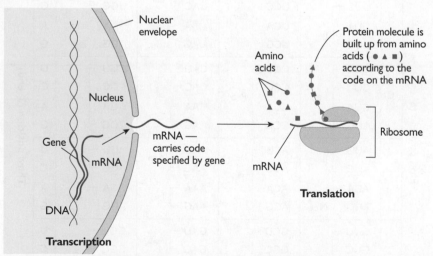

Figure 45 An overview of protein synthesis

Protein synthesis consists of the sequence of events shown in Figure 45.
- The DNA base sequence is copied as a molecule of **messenger RNA** (mRNA); this process is called **transcription**.
- The mRNA travels from the nucleus through pores in the nuclear envelope to the ribosomes.

- Free amino acids are carried from the cytoplasm to the ribosomes by molecules of **transfer RNA** (tRNA).
- Ribosomes 'read' the mRNA code and assemble the amino acids into a protein; this process is called **translation**.

During transcription, the non-coding (antisense) strand of the DNA is used as the template to synthesise the mRNA. As a result, the mRNA has a base sequence that is *complementary* to the antisense strand of the DNA and, therefore, is the *same* as the sense strand (except that uracil replaces thymine).

The genetic code is:

- a triplet code — three bases code for one amino acid. In DNA, the three bases coding for one amino acid make up a **triplet**. In mRNA, the three bases coding for one amino acid make up a **codon**.
- degenerate — there are more 'codes' than there are amino acids — 64 triplets and only 20 amino acids. Some amino acids have more than one code and some DNA triplets say 'stop transcribing' (some RNA triplets say 'stop translating').
- non-overlapping — the three bases that form one triplet are not part of any other adjacent triplet
- universal — the codes are the same in all organisms

The table shows the mRNA codons for the 20 amino acids used in protein synthesis. It also shows the 'stop translation' codes.

	Second position				
First position (5' end)	**U**	**C**	**A**	**G**	Third position (3' end)
U	UUU ⌉ Phe UUC ⌋ UUA ⌉ Leu UUG ⌋	UCU ⌉ UCC UCA Ser UCG ⌋	UAU ⌉ Tyr UAC ⌋ UAA stop UAG stop	UGU ⌉ Cys UGC ⌋ UGA stop UGG Trp	U C A G
C	CUU ⌉ CUC CUA Leu CUG ⌋	CCU ⌉ CCC CCA Pro CCG ⌋	CAU ⌉ His CAC ⌋ CAA ⌉ Gln CAG ⌋	CGU ⌉ CGC CGA Arg CGG ⌋	U C A G
A	AUU ⌉ Ile AUC AUA ⌋ AUG Met	ACU ⌉ ACC ACA Thr ACG ⌋	AAU ⌉ Asn AAC ⌋ AAA ⌉ Lys AAG ⌋	AGU ⌉ Ser AGC ⌋ AGA ⌉ Arg AGG ⌋	U C A G
G	GUU ⌉ GUC GUA Val GUG ⌋	GCU ⌉ GCC GCA Ala GCG ⌋	GAU ⌉ Asp GAC ⌋ GAA ⌉ Glu GAG ⌋	GGU ⌉ GGC GGA Gly GGG ⌋	U C A G

Key facts you must know and understand

Transcription of DNA to messenger RNA

Compared with DNA, a molecule of mRNA:

- is shorter
- is single-stranded (not double-stranded)
- contains the pentose ribose in place of deoxyribose
- contains the base uracil in place of thymine

Transcription (Figure 46) takes place in the following way:

- The enzyme **RNA polymerase** binds with a section of DNA next to the gene that is to be transcribed.
- Transcription factors (see page 56) activate the enzyme.
- RNA polymerase breaks the hydrogen bonds between the strands of DNA in the region that makes up the gene and moves along the antisense strand, using it as a template for synthesising mRNA.
- RNA polymerase assembles free RNA nucleotides into a chain in which the base sequence is complementary to the base sequence on the antisense strand of the DNA.
- The completed molecule leaves the DNA; the strands of DNA rejoin and re-coil.

The mRNA molecule now contains transcripts of:

- the **exons** (regions that code for amino acids)
- the **introns** (non-coding regions that separate the coding regions)

Transcription factors (see page 56)

Examiner tip

Candidates are often confused about terms such as sense/antisense strands and coding/non-coding strands. As long as you make it clear that only one of the two DNA strands is transcribed, an examiner will reward you — even if you use neither set of terms.

Examiner tip

Be sure to write that RNA polymerase assembles RNA *nucleotides*. Weaker candidates tell us that they assemble bases.

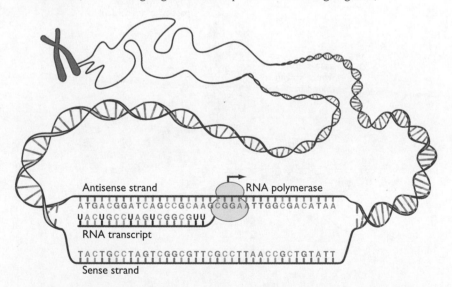

Figure 46 Transcription — the base sequence of the mRNA is identical to the sense strand of DNA, except that thymine (T) is replaced by uracil (U)

At this stage, the transcribed molecule is referred to as **pre-mRNA**.

splicing

Next, the introns are 'cut out' and the remaining exons are spliced together. The triplets of bases in mRNA are called **codons**.

The production of functional mRNA from pre-RNA is shown in Figure 47.

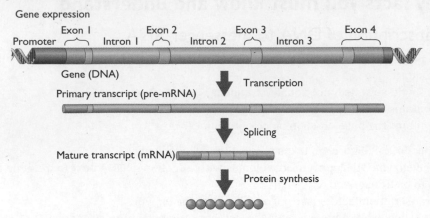

Figure 47 Producing functional mRNA from pre-mRNA

Translation of messenger RNA to a polypeptide

The structure of transfer RNA (tRNA)

Each tRNA molecule carries an amino acid to a ribosome. Each of the different types of tRNA has a structure that allows it to transfer just one specific amino acid *and* to be recognised as carrying that amino acid (and no other).

If asked to compare the structures of mRNA and tRNA, do not write that tRNA is double-stranded. Both consist of a single strand of nucleotides.

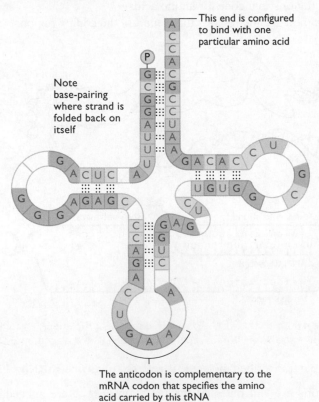

Figure 48 The structure of tRNA

However, all tRNA molecules have the same basic structure, with the two key features shown in Figure 48:

- On one part of the molecule is a triplet of bases called an **anticodon** with a base sequence complementary to one of the mRNA codons.
- At one end of the tRNA molecule there is an **attachment site** for the amino acid specified by the mRNA codon.

The stages in translation of mRNA into a polypeptide

Translation is shown in Figure 49.

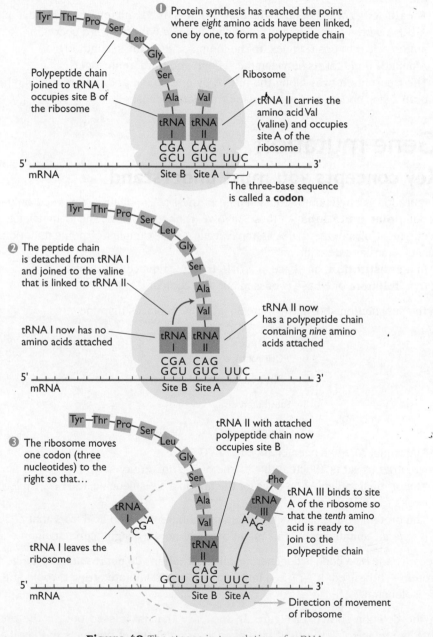

① Protein synthesis has reached the point where *eight* amino acids have been linked, one by one, to form a polypeptide chain

Polypeptide chain joined to tRNA I occupies site B of the ribosome

Ribosome

tRNA II carries the amino acid Val (valine) and occupies site A of the ribosome

The three-base sequence is called a **codon**

② The peptide chain is detached from tRNA I and joined to the valine that is linked to tRNA II

tRNA I now has no amino acids attached

tRNA II now has a polypeptide chain containing *nine* amino acids attached

③ The ribosome moves one codon (three nucleotides) to the right so that...

tRNA II with attached polypeptide chain now occupies site B

tRNA III binds to site A of the ribosome so that the *tenth* amino acid is ready to join to the polypeptide chain

tRNA I leaves the ribosome

Direction of movement of ribosome

Figure 49 The stages in translation of mRNA

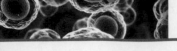

The following sequence of events follows the attachment of a ribosome to the end of a molecule of mRNA:

- The first two codons of the mRNA enter the ribosome. (Translation always begins at the first AUG codon in the mRNA sequence.)
- Transfer RNA molecules (with amino acids attached), that have complementary anticodons to the first two codons of the mRNA, enter the ribosome and bind to the respective codons.
- A peptide bond forms between the amino acids carried by the two tRNA molecules.
- The ribosome moves along the mRNA by one codon, bringing the third codon into the ribosome.
- The tRNA that was bound to the first codon is freed and returns to the cytoplasm.
- A tRNA with a complementary anticodon binds with the third codon, bringing its amino acid into position next to the amino acid on the second tRNA.
- A peptide bond forms between the second and third amino acids.
- The ribosome moves along the mRNA by one codon and the process is repeated until a stop codon is in position and translation ceases.

Knowledge check 45

A molecule of tRNA is attracted to a codon that is GAU. (a) What is the anticodon on the tRNA? (b) What amino acid is the tRNA molecule carrying?

Gene mutation

Key concepts you must understand

A mutation is any spontaneous change in the DNA molecule. You need only know about **point mutations** — these involve changes to just one DNA triplet. They occur most often when DNA is replicating. Two examples of point mutations are substitutions and deletions.

- In a **substitution**, one base in a DNA triplet is replaced by another.
- In a **deletion**, one base is missed out (not copied).

In the substitution shown in Figure 50, guanine replaces thymine:

A-T-T -T-C-C -G-T-T -A-T-C ...
↑
Original base

A-T-G -T-C-C -G-T-T -A-T-C ...
↑
Substituted base

Figure 50 Base substitution

- The triplet ATT has been changed to ATG.
- No other triplet is affected (this is true of all substitutions).
- The original triplet, ATT, codes for the amino acid isoleucine; the new triplet, ATG, codes for methionine.
- The protein synthesised will contain one different amino acid from that coded for by the un-mutated gene; this might or might not be a significant change.

Knowledge check 46

Explain why the different amino acid in the protein produced following a point mutation might be 'a significant change'.

Because the DNA code is degenerate, all substitutions do not result in a change in the protein synthesised. If ATT had mutated to ATC, the mutated triplet would still code for isoleucine.

In the deletion shown in Figure 51, the base that is 'missed out' in the mutated DNA is replaced, as the first base of each original triplet becomes the last base of the preceding triplet. This is called a **frame shift** and produces a totally new base sequence that results in a nonsense code and either a non-functional protein or no protein at all.

A-T-T -T-C-C -G-T-T -A-T-C ...
↑
Deletion here

A-T-T -C-C-G -T-T-A -T-C ...
↑
Replaced by first base of next triplet

Figure 51 Base deletion

In the example in Figure 51, the first triplet remains the same because the first base in the second triplet happens to be the same as the deleted third base in the first triplet. All the triplets after the deletion are altered.

Mutations are not passed on to the next generation unless they occur in either:
- a sex cell, or
- a cell in the sex organ that divides to form the sex cells

Key facts you must know and understand

Mutations occur spontaneously and randomly. However, the rate of mutation can be increased by a number of factors including:
- carcinogenic chemicals — for example benzene in tobacco smoke
- high-energy radiation — for example ultraviolet radiation, X-rays

A mutation in a normal body cell could have one of several consequences. It could:
- be completely harmless
- damage the cell
- kill the cell
- make the cell cancerous, which might kill the person

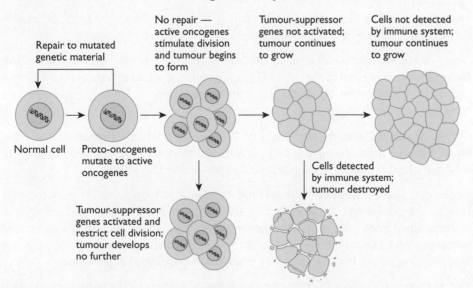

Figure 52 How mutations may result in tumour formation

S Knowledge check 47

(a) How might a tumour damage the body? (b) How does a malignant tumour (cancer) differ from a benign tumour?

Proto-oncogenes and **tumour suppressor genes** play important roles in regulating cell division and preventing the formation of a tumour. The flowchart in Figure 52 shows how mutations in these genes, if they are not repaired, can result in tumour formation.

Summary

After completing this topic, you should be able to:

- explain why the genetic code is described as degenerate, non-overlapping and universal
- compare the structure and composition of DNA, mRNA and tRNA
- describe the processes of transcription and translation during polypeptide synthesis
- explain the difference in the origin and consequence of deletion and substitution gene mutations
- describe the effect of mutations in proto-oncogenes and tumour-suppressor genes on the rate of cell division
- interpret the sequences of nucleic acid bases and amino acids when provided with suitable data
- intepret data from experimental work on the role of nucleic acids

The control of gene action

Key concepts you must understand

In most cells, only a small fraction of the genes are active. Different genes are active in different types of cell and also at different times in the same cell.

Not all DNA codes for proteins. Non-coding DNA includes:
- introns within genes
- minisatellite and microsatellite regions between genes
- **promoter** and **enhancer** regions of DNA that are close to the coding DNA to which RNA polymerase binds when it initiates transcription of that gene

Promoter and enhancer regions of DNA can also be switched 'on' and 'off', otherwise the regulated gene would be permanently active or permanently inactive.

Genes can be actively 'switched on' in the following way. Proteins called **transcription factors** that are present in the cell bind with the promoter and enhancer regions. RNA polymerase is then able to recognise the 'promoter/transcription factor complex' and transcribe the gene (Figure 53).

Genes can also be actively 'switched off'. This happens when a **repressor** molecule binds with the promoter region and prevents transcription factors from binding. In this state, RNA polymerase cannot bind and transcribe the gene.

Genes can also be '**silenced**'. This is not the same as repressing the action of the gene as the gene is still active. Cells can 'silence' a gene by degrading the mRNA that is transcribed from it. So, although the gene is still active, no protein is synthesised because no mRNA reaches the ribosomes to allow translation to occur. Gene silencing by degrading mRNA is one example of **post-transcriptional repression**.

The reason why different cells express different genes is because they contain different transcription factors and different repressors.

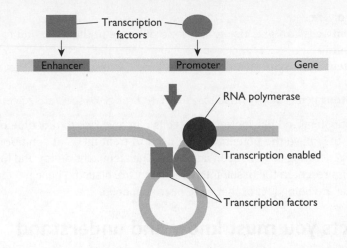

Knowledge check 48
Use Figure 53 to suggest why promoter and enhancer regions are described as *upstream* of the gene they regulate.

Figure 53 Transcription factors allow RNA polymerase to bind to, and transcribe, a gene

Some cells have the potential to express all their genes. These cells are called **totipotent cells** and are:

- unspecialised (they have not differentiated for any particular function as has, say, a neurone)
- capable of dividing and renewing themselves for long periods
- capable of giving rise to specialised cells

Specialisation of cells during development occurs by genes being repressed. The only genes that are capable of being expressed in a neurone are those that are needed for that particular function — the gene for brown pigmentation of the iris (for example) has been repressed.

Unspecialised cells that divide to give specialised cells are called **stem cells**. There is a hierarchy of specialisation. The hierarchy in humans is shown in Figure 54.

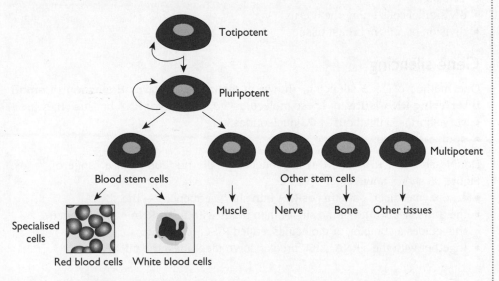

Figure 54 The hierarchy of cell specialisation in humans

Bone marrow transplants are sometimes given to help people produce functional white blood cells. Suggest what type of cell — totipotent, pluripotent or multipotent — these transplants will contain. Explain your answer.

S Knowledge check 50

Use your understanding of biological principles to suggest why transcription factors enable RNA polymerase to bind to a promoter region.

When they divide:

- **totipotent** cells can give rise to any kind of tissue in the body and to the cells of the placenta
- **pluripotent** cells can give rise to any kind of tissue but not the cells of the placenta
- **multipotent** cells can give rise to a restricted range of specialised cells

Because totipotent and pluripotent stem cells can give rise to any type of cell in the adult body, they have the potential to be used to treat medical conditions in which cells are degenerating or have been damaged. The stem cells divide and form the type of cell that is present in the tissue into which they are placed. These new cells replace the cells that are being lost or that have been damaged.

Key facts you must know and understand

Transcription factors

One model of the way in which transcription factors regulate gene action is as follows:

- The transcription factor binds to a **promoter sequence** of DNA upstream of the gene to be activated.
- RNA polymerase binds to the promoter sequence–transcription factor complex.
- RNA polymerase is 'activated' and moves away from the DNA–transcription factor complex along the DNA of the coding gene.
- The RNA polymerase transcribes the antisense strand of the DNA as it moves along; the gene is now being expressed.

The hormone oestrogen binds with receptors in certain cells to form an oestrogen–receptor complex. This complex acts as a transcription factor and binds to promoter regions of genes that stimulate cell division. This allows the hormone to initiate such effects as:

- division of cells lining the uterus
- division of cells in breast tissue

Gene silencing

One method of gene silencing, shown in Figure 55, involves the action of **small interfering RNA** (**siRNA**). These molecules are unusual for RNA because they are:

- very short — only about 21–23 nucleotides long
- double-stranded

Double-stranded RNA (dsRNA) is produced in the nucleus from a range of genes. Figure 55 shows how:

- an enzyme 'dicer' cuts this dsRNA into short sequences — the siRNA
- the antisense strand of the siRNA then binds with the mRNA it is to silence
- this guides a complex of molecules, called RISC, to the site
- together with the siRNA, RISC breaks down (degrades) the mRNA

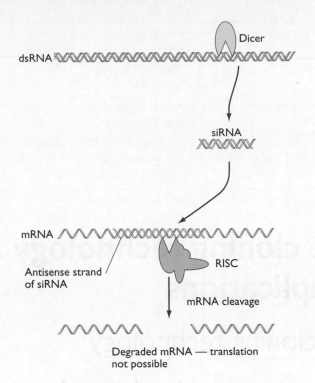

Figure 55 Gene silencing by siRNA

Knowledge check 51

Is a silenced gene (a) being transcribed, (b) being translated?

Totipotent cells in plants

More cells in plants are either totipotent or pluripotent than is the case in animals. Because of this, mature plants can be grown from small parts of plants called **cuttings**. One way to take a cutting is as follows:

- Cut a small side shoot from a plant.
- Remove some leaves (to prevent excessive water loss).
- Place the cutting in compost and keep the compost watered.

The base of the cutting develops small roots and, eventually, a full root system. This means that the cells at the base of the cutting must have started to divide again and the cells they formed then specialised into the various tissues in the roots. To do this, they must have been at least pluripotent.

In micropropagation, the totipotency or pluripotency of some plant cells is exploited even further. Small sections of plant tissue called **explants** are grown in special culture media. The explants are treated first with plant growth factors that make them develop roots and later with other plant growth factors that make them develop shoots. When they have developed sufficiently, the young plantlets are transplanted into compost and grown in glasshouses.

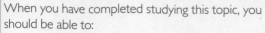

Summary

When you have completed studying this topic, you should be able to:

- explain the difference between totipotent, pluripotent and multipotent cells
- describe how the presence of totipotent cells in plants can be used in plant propagation
- suggest how totipotent cells in humans, called stem cells, could be used to treat genetic disorders, and evaluate their use in humans
- describe the role of transcriptional factors, repressors and promoter regions in the control of gene expression
- describe the role of siRNA in gene silencing
- interpret information about the control of gene expression
- evaluate unemotionally the use of gene control in the treatment of disorders resulting from hereditary, or acquired, gene mutations

Gene cloning technology and its applications

Gene cloning technology

Key concepts you must understand

Cloning a gene produces multiple, identical copies of that gene. There are two main types of gene cloning:

- **in vivo cloning** — the gene is introduced into a cell and is copied as the cell divides
- **in vitro cloning** — DNA (containing the gene in question) is copied many times over by the **polymerase chain reaction** (**PCR**) in a PCR machine

Gene (DNA) probes can help to identify genes. A gene probe is a short length of single-stranded DNA that is tagged with a radioactive or, more commonly, fluorescent marker. The sequence of bases is complementary to at least part of the sequence of bases in the gene of interest. When a gene probe is mixed with a sample of single-stranded DNA containing the gene for which it is specific, the gene probe hybridises with the DNA. The marker enables us to see that this has happened.

Once identified, the gene to be cloned can be obtained by:
- extracting it from a donor cell using enzyme technology
- creating it in vitro from the corresponding mRNA

Once the gene has been cloned, it can be:
- stored in a **gene library**
- transferred into another cell, often from a different species, which, as a result, becomes **transgenic** (has a gene from a different species)

Vectors (carriers) are needed to transfer DNA into another cell. There are two types of vector:
- **plasmids** (small circular pieces of DNA found in some bacteria and yeasts)
- **viruses** that have been modified to be non-pathogenic

Knowledge check 52

(a) What is a gene probe?
(b) How might it be used?

Examiner tip

The term *vector* has more than one scientific meaning. In gene cloning technology, it refers specifically to a carrier of DNA from one cell to another.

Gene sequencing is a technique that allows biologists to determine the sequence of bases in a section of DNA.

The first gene sequencing technique was the **chain terminator technique** developed by Fred Sanger. This technique works by:
- creating many sections of DNA complementary to part of one strand of the DNA being investigated
- ensuring that some sections are just one nucleotide long, some two, some three and so on all the way up to sections that are 'full length'
- finding out which base is at the end of a strand that is one nucleotide long, which base is at the end of a strand that is two nucleotides long and so on up to the 'full length' DNA

The sequence of bases at the ends of the newly created strands is complementary to the sequence in the DNA being investigated.

Many gene sequencing techniques can only determine the base sequence of relatively small sections of DNA. To determine the base sequence of larger sections, gene sequencing is sometimes combined with **restriction mapping**. This technique uses **restriction enzymes**. These enzymes cut DNA at specific **restriction sites** — specific base sequences, for example CTTAAG. The size of the DNA fragments produced can be determined by gel electrophoresis.

Gel electrophoresis is a technique used to separate charged particles. Negatively charged ions move through a gel towards a positive electrode. The gel acts as a 'molecular sieve', slowing down larger particles which, therefore, do not move as far as smaller ones.

Using two (or more) restriction enzymes together with gel electrophoresis allows us to make predictions about possible locations of their restriction sites in the DNA.

Using gel electrophoresis to check the size of the fragments produced by treatment with both restriction enzymes simultaneously will confirm one of the predictions (see Figure 56).

Knowledge check 53

Explain why a restriction endonuclease will only cut DNA at a specific restriction site.

Knowledge check 54

Which part of a DNA fragment carries a negative charge?

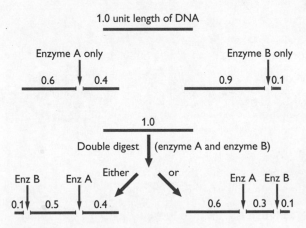

Figure 56 Predicting the location of restriction sites in DNA

Figure 57 shows that, to sequence a really large section of DNA, we could:
- digest the large piece of DNA into several smaller pieces

- construct a restriction map for each piece
- compare these restriction maps and see if there was any overlap of the restriction sites, which would suggest a common piece of DNA

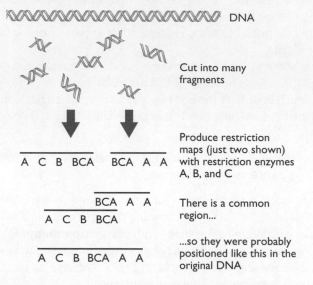

Figure 57 Using restriction mapping to determine base sequences of large fragments of DNA

Knowledge check 55

A restriction enzyme cuts a sample of DNA into six fragments. How many times did the restriction site occur in the original DNA (a) if the DNA was linear, (b) if the DNA was circular?

Once the DNA fragments have been sequenced, we can now assemble the base sequences into a larger sequence, remembering not to duplicate the overlaps.

Key facts you must know and understand

In vivo gene cloning

The flow chart in Figure 58 shows the main stages of in vivo gene cloning.

Knowledge check 56

Figure 58 shows two ways of obtaining a gene. Suggest *one* advantage of manufacturing a gene from mRNA rather than removing the gene from a cell.

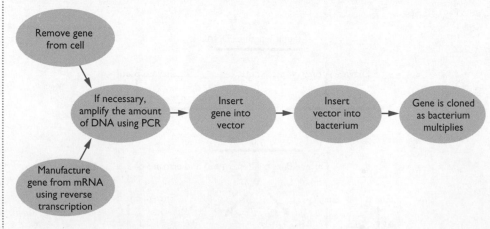

Figure 58 The main stages of in vivo gene cloning

Removing the gene from the donor cell

The donor cells are incubated with restriction enzymes whose restriction sites will ensure that a section of DNA containing the gene is isolated.

Some restriction enzymes do not make a 'clean cut' across the two strands of the DNA, but make a staggered cut, leaving unpaired bases. These staggered ends, shown in Figure 59, are often called '**sticky ends**'.

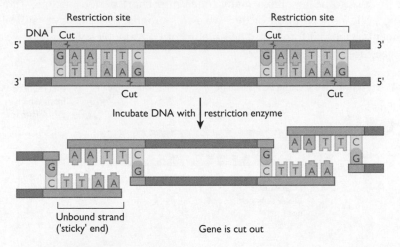

Figure 59 Staggered cuts in DNA produce 'sticky ends'

Knowledge check 57

What is the advantage of using restriction enzymes that produce 'sticky ends'?

Creating the gene from mRNA

This technique uses an enzyme called **reverse transcriptase**. This enzyme reverses the process of transcribing DNA into mRNA and transcribes mRNA into DNA.

Figure 60 shows the following main stages of the process:
- Incubate the relevant mRNA molecule with reverse transcriptase and the necessary free DNA nucleotides.
- Reverse transcriptase creates a single strand of complementary DNA (cDNA).
- 'Wash out' the mRNA.
- Incubate the cDNA with DNA polymerase and free DNA nucleotides.
- DNA polymerase creates a complementary strand of DNA which bonds with the strand created by reverse transcriptase.

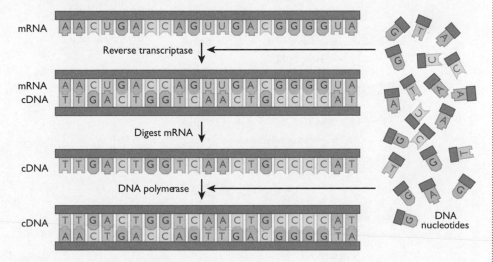

Figure 60 Creating the gene using reverse transcriptase

Amplifying the amount of DNA using PCR

The polymerase chain reaction (PCR) is an automated technique that allows a tiny sample of DNA to be amplified many times in a short period of time. There is a repeating cycle of separation of the two DNA strands, followed by synthesis of a complementary strand for each. The amount of DNA doubles with each cycle. The main stages are shown in Figure 61.

Knowledge check 58

It is important that the DNA sample amplified by the PCR is not contaminated. Explain why.

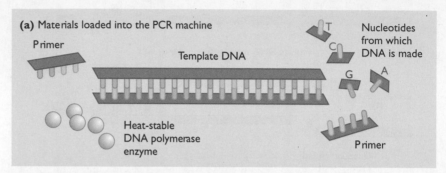

(a) Materials loaded into the PCR machine

Primer

Template DNA

T
C
G
A
Nucleotides from which DNA is made

Heat-stable DNA polymerase enzyme

Primer

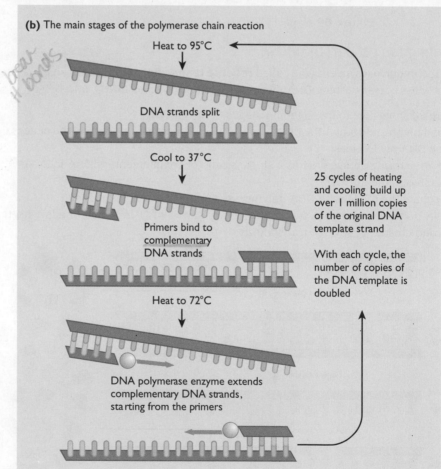

(b) The main stages of the polymerase chain reaction

Heat to 95°C

DNA strands split

Cool to 37°C

Primers bind to complementary DNA strands

Heat to 72°C

DNA polymerase enzyme extends complementary DNA strands, starting from the primers

25 cycles of heating and cooling build up over 1 million copies of the original DNA template strand

With each cycle, the number of copies of the DNA template is doubled

Knowledge check 59

What is the importance of the primer in Figure 61?

Figure 61 The polymerase chain reaction

The DNA polymerase used is thermostable so it can withstand the temperature of 95°C. It has an optimum temperature of 72°C, which is why the replication phase is carried out at that temperature.

Transferring the gene into a vector

Plasmids and viruses have both been used as vectors, but plasmids are usually preferred. Figure 62 shows how a gene is transferred into a plasmid.

- The plasmid is cut open using the same restriction enzyme used to cut out the DNA fragments from the donor cell. The 'sticky ends' of the two types of DNA will contain complementary base sequences (if the gene was synthesised from mRNA, 'sticky ends' are added to its blunt ends).
- The DNA fragments are incubated with the plasmids. The plasmid DNA and the gene DNA **anneal** (join) in the following way:
  - Hydrogen bonds form between the bases in the 'sticky ends', weakly holding the gene DNA in place in the plasmid.
  - Catalysed by the enzyme **ligase**, covalent bonds form between the sugar–phosphate backbones of the plasmid DNA and gene DNA; the gene has now been firmly **spliced** into the plasmid.

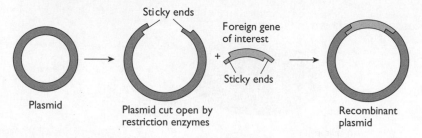

Figure 62 Transferring a gene into a plasmid

Any DNA that has had 'foreign DNA' inserted into it is called **recombinant DNA**, so the plasmid is now a **recombinant plasmid**.

Transferring the plasmid into a bacterium

The bacteria are treated with a solution of calcium chloride, which makes their cell walls permeable to plasmids. The bacteria are then incubated with the plasmids. However:

- the frequency of plasmid take-up by the bacteria may be as low as 1 in 10 000
- some bacteria will take up recombinant plasmids and others will take up the original, non-recombinant plasmids; those that take up the recombinant plasmids are called **transformed bacteria**

There are two main ways of checking which bacteria have taken up the recombinant plasmids.

Method 1: using marker genes

Some plasmids contain genes that confer resistance to two antibiotics — for example, both ampicillin *and* tetracycline. Introducing a gene into such a plasmid (Figure 63) splits the gene for tetracycline resistance and makes it inactive. At the end of the process, there will be the following types of bacteria:

Knowledge check 60

Define the terms *recombinant* and *transgenic*.

- those that have not taken up any plasmids and are not resistant to either antibiotic
- those that have taken up the original plasmids and are resistant to both antibiotics
- those that have taken up the recombinant plasmids and are resistant to ampicillin only

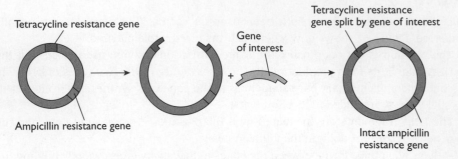

Figure 63 Inserting a gene into a plasmid can split a gene already present

The bacteria are cultured on media containing ampicillin or tetracycline. Those that survive on the ampicillin culture *only* are the transformed bacteria.

Method 2: using DNA probes

This technique is shown in Figure 64.

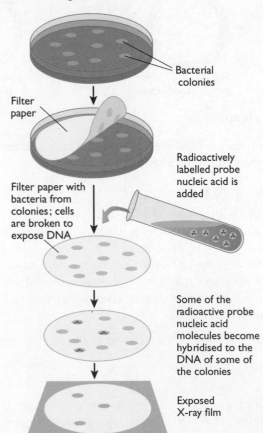

Knowledge check 61

What percentage of bacteria in Figure 64 were transformed?

Figure 64 Using a DNA probe to identify transformed bacteria

AQA A2 Biology

- Culture the bacteria on agar gel in Petri dishes. Each bacterium will occupy a unique location on the surface of the agar.
- Incubate the Petri dish at a suitable temperature so that each bacterium multiplies to form a colony of millions (all containing copies of any plasmid taken up by the original bacterium).
- 'Blot' each Petri dish with filter paper to transfer a few cells from each colony to the filter paper in the same relative positions as the colonies in the Petri dish.
- Break open the cells in each sample on the filter paper and split the two strands of DNA.
- Incubate with a radioactive gene probe (with a sequence complementary to the gene of interest). Only DNA from cells with the gene (in the recombinant plasmid) will bind with the probe and become radioactive.
- Locate the radioactive DNA on the filter paper by producing an X-ray photograph.
- The positions of the radioactive DNA on the X-ray photograph correspond to those colonies of bacteria in the Petri dish that contain the gene of interest.

Once the bacteria that contain the gene have been identified, they can be cultured to give large numbers of bacteria for a specific purpose or simply to clone the gene.

DNA sequencing: the chain terminator technique

This technique has some parallels with PCR, as it also requires:
- a single-stranded section of DNA to act as a template
- DNA polymerase
- DNA primers
- DNA nucleotides

The single-stranded DNA is primed and mixed with DNA polymerase, DNA nucleotides and specially modified DNA nucleotides called **dideoxynucleotides** (ddNTPs). A ddNTP can form a sugar–phosphate bond with only one other nucleotide. If it enters a growing chain of DNA, the chain cannot be extended any further because the ddNTP has already formed the only sugar–phosphate bond it is able to make.

These are the events that are involved in the chain terminator technique:
- Four samples of the single-stranded DNA are primed. In Figure 65, the primer is single-stranded DNA with the base sequence GTC.
- Each sample is mixed with all four normal nucleotides and one dideoxynucleotide (e.g. dideoxycytosine (ddC) or dideoxythymine (ddT)), which is either radioactive or tagged with a fluorescent dye.
- DNA polymerase is added and it begins to build a chain complementary to the single-stranded DNA sample.
- In the dideoxycytosine (ddC) reaction tube, polymerase uses ddC in the first available position (opposite guanine) on some of the strands, which then stop building. All the other strands continue building.
- In some of the other DNA strands, normal cytosine is used in the first position and ddC is used in the second available position. These strands then stop building.
- In yet other DNA strands, the ddC is not used until the third available position, in others the fourth available position...and so on.
- At the end of the run in that tube, there will be new DNA strands with ddC in each available position. The ddC nucleotide will be at the end of each of these strands.

- A similar process will occur in the other three reaction tubes, but with different ddNTPs at the end of the fragments.

The fragments from each tube are then separated at the same rate for the same length of time by gel electrophoresis. Figure 65 shows how strands of different lengths from the different tubes might be distributed. The different fragments move different distances, according to their masses. In this sample, the shortest fragment ended ddA and so moved furthest. The longest also ended ddA and moved least. All the strands begin with CTG because this is the region that is complementary to the primer on the DNA.

Knowledge check 62

What was the sequence of bases in the *original* piece of DNA used in Figure 65?

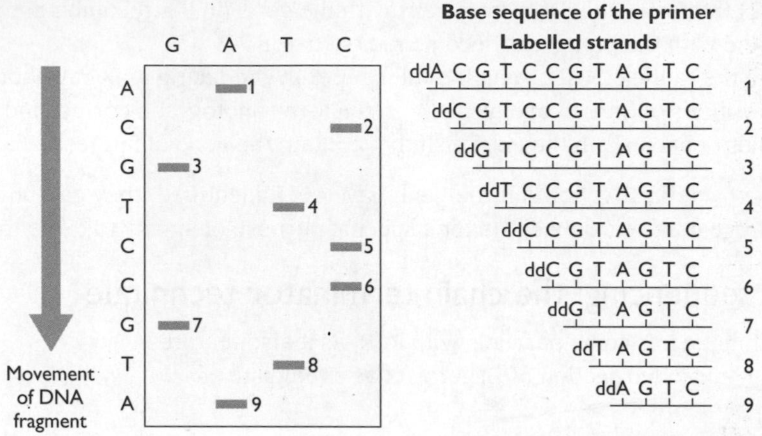

Figure 65 Results of gel electrophoresis of DNA fragments produced by the chain terminator technique

Applications of gene cloning technologies

Key facts you must know and understand

Genetically modified organisms are used to manufacture specific products that can be injected into humans (e.g. insulin, bovine somatotrophin and some vaccines). Other organisms have been genetically modified to produce increased yields (e.g. of cereals, or of milk in cattle).

Gene technology is also used to produce **DNA microarrays** (**DNA chips**) which allow the identification of, and assessment of the activity of, specific genes. A microarray contains thousands of gene probes, each specific for a particular gene. These react with a person's genes to produce a certain colour. The position of the probe in the microarray identifies the gene; the tone and the intensity of the colour allow computers to analyse the activity of the gene.

Gene sequencing and restriction mapping can identify human genes and provide information for genetic counsellors to advise patients on what options may be open to them, knowing that they have certain defective genes and may pass these genes on to their children.

Genetic fingerprinting is a technique for comparing the non-coding DNA of different people. The coding DNA within the genes does not vary a great deal between individuals — for example, the base sequence in the gene for normal haemoglobin is the same in nearly everyone. Non-coding DNA does vary a great deal and is inherited along with the coding DNA. Genetic fingerprints can, therefore, be used to help resolve disputed parentage — each fragment of DNA in the genetic fingerprint of a child must have come from one parent or the other. Genetic fingerprints are prepared as shown in Figure 66.

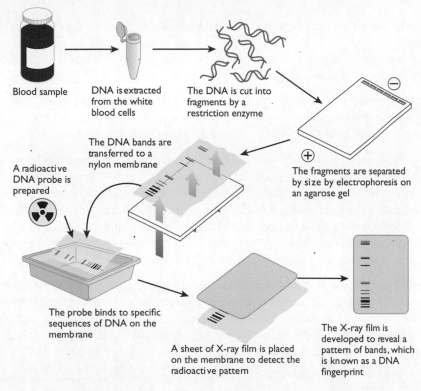

Blood sample

DNA is extracted from the white blood cells

The DNA is cut into fragments by a restriction enzyme

The DNA bands are transferred to a nylon membrane

A radioactive DNA probe is prepared

The fragments are separated by size by electrophoresis on an agarose gel

The probe binds to specific sequences of DNA on the membrane

A sheet of X-ray film is placed on the membrane to detect the radioactive pattern

The X-ray film is developed to reveal a pattern of bands, which is known as a DNA fingerprint

Figure 66 The main stages in obtaining a genetic fingerprint

Knowledge check 63

Non-coding DNA is useful in genetic fingerprinting. Explain why.

Gene therapy involves treating genes that cause disease — for example in cystic fibrosis and in sickle-cell anaemia. Because these genes act only in some cells of the body, the cells in which they act can be targeted. Researchers in this area use three principal techniques:

- Add a normal gene to the affected cells to restore normal functioning.
- Repair the abnormal gene by selective reverse mutation.
- Regulate the activity of the gene (it could be 'silenced' or 'turned on' as required).

Much of the research to date has centred on the first technique, often using viruses as vectors for the gene. Progress has been limited because of:

- the low rate of gene take-up by cells with the defective gene
- attack of the virus particles by the recipient's immune system
- the virus vectors regaining virulence and causing disease

There are moral and ethical concerns about the use of gene technology.

- **Morality** is our personal sense of what is right and what is wrong.
- **Ethics** represent the 'code' adopted by a particular group to govern its way of life.

The concerns that some people have about gene technology include the following:
- *A species is sacrosanct and should not be altered genetically in any way.*

People who take this moral stance usually do so on the basis that the genes from one species would not normally find their way into another species. However, genes have been 'jumping' from one species to another for millions of years.
- *Not enough is known about the long-term ecological effects of introducing genetically modified organisms into the field. They may out-compete wild plants and take over an area.*

The effects of any new crop cannot be determined without field trials. Does this make it wrong?
- *If plants are genetically engineered to be resistant to herbicides, the gene could 'jump' into populations of weeds and other wild plants.*

Herbicide-resistant strains of many plants already exist naturally. The gene could just as easily jump from these.
- *Gene technology might give doctors the ability to create designer babies.*

Most doctors would find this morally and ethically unacceptable. They might consider replacing genes that cause disease but not replacing genes merely to improve a child's image in the eyes of its parents. However, if such practices become possible, who will define for doctors what is ethically acceptable? What will be the dividing line between cosmetic gene therapy and medical gene therapy?
- *Using genetic fingerprinting to combat crime will only be useful if there is a genetic database — a file of the genetic fingerprints of everyone in the country.*

Who will have access to this information? If insurance companies had access to the genetic database, they might refuse insurance (or charge higher premiums) to people with an increased risk of, say, heart disease. Employers could (covertly) refuse employment to people because their 'genetic profiles' did not meet particular requirements. A recent ruling from the European Court of Human Rights states that the police have no right to hold the DNA of someone unless that person has been convicted of a crime.

Summary

After completing this topic, you should be able to:
- describe how fragments of DNA can be produced by converting mRNA to cDNA, by using restriction enzymes and by the polymerase chain reaction (PCR)
- describe, and compare the relative advantages of, in vivo and in vitro techniques for gene cloning
- describe the use of recombinant DNA technology to produce transformed organisms that are of benefit to humans
- interpret information about the use of recombinant DNA technology
- evaluate the moral, ethical and social issues associated with recombinant DNA technology
- describe the potential for, and evaluate the effectiveness of, gene therapy to supplement defective genes
- outline the use of DNA in screening, restriction mapping and DNA sequencing in medical diagnosis
- outline the biological principles that underpin genetic fingerprinting and explain the use of genetic fingerprints in forensic science, medical diagnosis and plant breeding
- interpret data showing the results of genetic fingerprinting

Questions & Answers

This section contains questions similar in style to those you can expect to see in the main body of **BIOL5**. The responses that are shown are real students' answers to the questions.

There are several ways of using this section. You could:

- 'hide' the answers to each question and try the question yourself. It needn't be a memory test — use your notes to see if you can actually make all the points you ought to make
- check your answers against the students' responses and make an estimate of the likely standard of your response to each question
- check your answers against the examiner's comments to see where you might have failed to gain marks
- check your answers against the terms used in the question — for example, did you *explain* when you were asked to, or did you merely *describe*?

Examiner's comments

Each question is followed by a brief analysis of what to watch out for when answering the question (shown by the icon ⓔ). All student responses are then followed by examiner's comments. These are preceded by the icon ⓔ and indicate where credit is due. In the weaker answers, they also point out areas for improvement, specific problems, and common errors such as lack of clarity, weak or non-existent development, irrelevance, misinterpretation of the question and mistaken meanings of terms

Tips for answering questions

Use the mark allocation. Except for the essay, one mark is allocated for one fact, concept or item in an explanation. Make sure your answer reflects the number of marks available.

Respond appropriately to the command words in each question, i.e. the verb the examiner uses. The terms most commonly used are explained below.

- **Describe** — in a question testing AO1, this means 'write what you have learnt about'; in a question testing AO2 it means 'translate data into words'.
- **Explain** — give biological reasons for *why* or *how* something happens.
- **Calculate** — do some kind of sum and show how you got your answer.
- **Compare** — show how the same property of organisms/processes is similar or different.
- **Complete** — add to a diagram, graph, flow chart or table.
- **Name** — rarely found in BIOL5, it literally means give the name of something.
- **Suggest** — give a plausible biological explanation for something; this term is often used in questions testing understanding in an unfamiliar context, which is common in BIOL5.
- **Use** — you must find in the question, and include in your answer, relevant information or data.

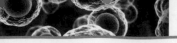

Question 1 Hormone action

The level of plasma glucose at rest is maintained largely by the hormones insulin and glucagon. These are both secreted by islet cells in the pancreas and affect mainly skeletal muscle cells and liver cells.

(a) Explain why the hormones affect mainly skeletal muscle cells and liver cells.   (2 marks)

(b) Explain how one molecule of glucagon can bring about the conversion of many molecules of glycogen to glucose.   (2 marks)

(c) Name one other hormone that can influence the level of plasma glucose.   (1 mark)

Total: 5 marks

ⓔ Notice that (a) and (b) require explanations, so you must give reasons, not descriptions. They also both test aspects of synopsis — your ability to use understanding from previous units. You should be able to answer part (c) in a single word.

Student A

(a) Liver and skeletal muscle cells are the target cells for insulin and glucagon so they bind to these cells **a** in particular.

(b) The hormone causes the cell to break down many molecules of glycogen, one after the other **b**.

(c) Adrenaline **c**

ⓔ **1/5 marks awarded** **a** This student uses the idea of binding, but does not state to what and does not attempt to explain specificity, so scores no marks. **b** This statement presents the same information that is given in the question, and so fails to score. You must beware of simply re-wording a question. **c** This answer is correct and scores 1 mark.

Student B

(a) Insulin and glucagon bind to protein receptors **a** in the plasma membranes of these cells. The protein receptors are shaped so that only these hormones will fit — other cells don't have the same **b** receptors.

(b) When one molecule of glucagon binds, it activates an enzyme **c** which catalyses the conversion of glycogen to glucose. An enzyme can catalyse the breakdown of many molecules of glycogen **c**. As soon as one leaves the active site of the enzyme, another one can enter to be broken down. This is the turnover rate of the enzyme.

(c) Thyroxine **d** increases the basal metabolic rate (BMR), increasing the rate at which glucose is used up.

ⓔ **5/5 marks awarded** This student has understood that, to answer this synoptic question, you need an appreciation of shape of protein receptor molecules from Unit 1. S/he has told us where the hormones bind **a** and explained why only cells with these receptors are affected **b**, scoring 2 marks. In (b), s/he clearly understands the cascade principle **c** of hormone action, and gains both marks. **d** S/he gains the available mark for this word; the rest of the answer, though true, was a waste of time. When asked to name, just write the name.

ⓔ This question is not just about the effects of hormones. It is about how they bring about their effects and, consequently, requires understanding of several areas from the specification. Student A should have been able to do better than 1 mark (grade U): the first section concerning targeting specific cells is straightforward if you understand the concept of specific receptor proteins. Student B scores all 5 marks (grade A).

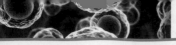

Question 2 **Action potentials**

Nerve impulses are propagated along the axons of neurones as a series of action potentials.

Figure 1 shows the changes in membrane potential, sodium ion (Na⁺) conductance and potassium ion (K⁺) conductance of an axon membrane as an action potential is generated.

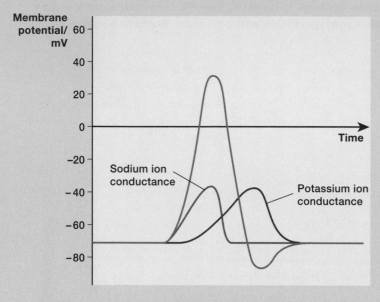

Figure 1

(a) Describe the evidence in the graphs that suggests that depolarisation is caused by an influx of sodium ions, while repolarisation is caused largely by the exit of potassium ions. (3 marks)

(b) Explain the role of the refractory period in the transmission of nerve impulses. (2 marks)

Total: 5 marks

ⓔ As you should expect in BIOL5, this graph contains unfamiliar information — the curves showing conductance. The examiner has given you sufficient information to enable you to use your understanding, though. Clearly, you should use information from the graph in answering part (a). By asking you to describe evidence, the examiner really does expect you to turn the relevant curves into words. Part (b) should involve fairly straightforward recall (AO1) but note that you are asked for an explanation, not a description.

Student A

(a) The increase in the membrane potential happens at the same time **a** as the increase in sodium conductance. This is when the action potential occurs, so it must be due to the sodium ions. When the membrane potential falls back to normal again, this corresponds with an increase in potassium conductance **a**.

(b) The refractory period is the period when an axon cannot have an action potential **b**. This is because it becomes permeable to sodium ions **c** and can't let them pass through.

ⓔ **2/5 marks awarded** **a** This student has linked the conductance curves to the curve showing membrane potential, so gains 1 mark. S/he has not linked the answer to depolarisation or repolarisation, as asked, and gains no further marks. **b** S/he gains 1 mark for the idea that action potentials cannot be generated **c** but then makes a contradictory statement by using the term 'permeable' to mean ions cannot pass through. An examiner will not choose which of two contradictory statements a candidate really means; if the answer is not clear, it gains no marks.

Student B

(a) The action potential is generated when the membrane potential becomes positive on the inside (it is usually –70 mV). The graph shows that this happens when the conductance to sodium ions increases, **a** allowing these ions to rush in. To restore the membrane potential to –70 mV, potassium ions rush out. This can only happen when the membrane becomes permeable to potassium, which is shown by the increase in potassium ion conductance, **a** after the peak occurs.

(b) This is the time when the membrane is impermeable to sodium and potassium ions, which means that a new action potential **b** cannot be generated.

ⓔ **3/5 marks awarded** Like student A, this student has not referred to depolarisation or repolarisation. **a** S/he has, however, given adequate descriptions of the events that are defined by these terms and has related the evidence to them, gaining 2 marks. **b** Like student A, s/he gains 1 mark for the idea that action potentials cannot be generated.

ⓔ **This topic involves concepts that many students find difficult, so make sure you *understand* it. You must also be confident in dealing with graphs so that you can relate unfamiliar information to more familiar information. Both students showed poor exam technique in part (a), by not using terms they were asked to use. Student A scores 2 marks (grade E) and student B scores 3 marks (grade D).**

Question 3 Protein synthesis

Protein synthesis takes place in the ribosomes. The code for synthesis of a particular protein is specified by a section of the **DNA** molecule and is carried to the ribosomes by **mRNA**.

(a) (i) What do we call a section of **DNA** that codes for a protein? (1 mark)

(ii) The **DNA** code is sometimes called a degenerate code. What does this mean? (1 mark)

(b) **Figure 1** shows protein synthesis taking place in a ribosome.

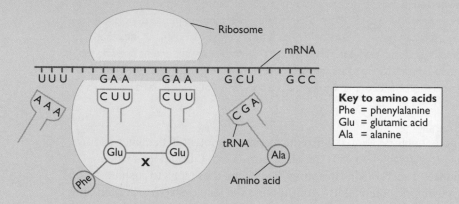

Figure 1

(i) Name the type of bond labelled **X**. (1 mark)

(ii) Use examples from the diagram to explain the terms *codon* and *anticodon*. (3 marks)

Total: 6 marks

ⓔ Part (a) is straightforward recall (AO1), which you should find easy. Make sure you actually use examples from the diagram in (b)(ii), otherwise you will not gain these easy 3 marks.

Student A

(a) (i) A gene **a**

(ii) There are more codes than there are amino acids **a**.

(b) (i) Glycosidic **b**

(ii) GAA is a codon **c**; CUU is an anticodon **c**

ⓔ **4/6 marks awarded a** These answers are correct and each gains 1 mark. **b** This answer is strange. Perhaps, despite the explanatory key to amino acids, this student saw 'Gly' on either side of bond **X** and became confused. **c** These two statements are correct and gain 2 of the 3 marks available. Although it can be inferred, this student has not made clear, for the third mark, that the codon and anticodon have complementary base sequences.

Student B

(a) (i)   A small section of DNA that codes for a protein is called a gene **a**.

   (ii)   Not all the triplet codes of the DNA code for amino acids **b**; some are stop
          codes.

(b) (i)   Peptide bond **c**

   (ii)   A codon is a triplet of bases on the mRNA molecule **d**.

ⓔ **4/6 marks awarded** **a** This answer is correct, but the word 'gene' was all that was required. **b** This answer gains the mark but, again, this student shows little appreciation of how to answer a question with a 1-mark tariff and wastes time writing more information than can be rewarded. **c** This answer is correct. **d** The student defines what a codon is and gains 1 mark. S/he has not used information from the diagram.

ⓔ **This is a topic that most candidates find easy — gaining marks is, largely, down to exam technique. Both students score 4 marks (grade C).**

Question 4 The polymerase chain reaction

The polymerase chain reaction (PCR) is a method of obtaining large amounts of DNA from a small initial sample. Figure 1 shows the main stages in the polymerase chain reaction.

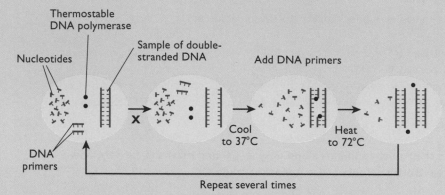

Figure 1

(a) (i)   What must be done to separate the strands of DNA (process X)?   (1 mark)

   (ii) What are primers?   (2 marks)

(b) (i)   What is meant by 'thermostable DNA polymerase'?   (2 marks)

   (ii) What is the main advantage of using a thermostable DNA polymerase in this process?   (1 mark)

Total: 6 marks

ⓔ If you understand this topic, you shouldn't really need the diagram to answer many of the questions. You do, however, need to use it in your answer to (a)(i) — look at the description of the final stage before answering it. Many students answer part (a)(ii) badly. You need to be able to give a precise statement.

Student A

(a) (i)   You must heat them **a**.

   (ii) Primers are small sections of DNA that start the new strands **b**.

(b) (i)   A thermostable DNA polymerase is one that is not easily denatured **c** at high temperatures.

   (ii) You can make the reaction faster by having it hotter **d**.

ⓔ **3/6 marks awarded** Although an examiner would not expect you to recall a precise temperature, s/he would expect you to use the diagram intelligently. You can see that the final stage is heated to 72°C, yet the DNA remains double-stranded. **a** So, an answer of 'heat them' is not enough. **b** In (a)(ii), the student seems to understand what a primer does but has failed to state that the DNA is single-stranded for the second mark. **c** The student has used a good A-level term in defining 'thermostable' but has not defined 'DNA polymerase', which was half the question. **d** This answer gains the mark.

Student B

(a) (i) The DNA must be heated to 95°C **a** to separate the strands.

(ii) Primers are small sections of single-stranded DNA **b** with a base sequence that is complementary to one end of the original DNA **c** strands.

(b) (i) A thermostable DNA polymerase has a tertiary structure which is more stable and less easily deformed **d** by heat.

(ii) The enzyme will still be active at high temperatures **e**. Probably its optimum temperature is high — it isn't easily denatured because of the strong bonds holding its tertiary structure.

ⓔ **4/6 marks awarded a** The student has actually given the correct temperature but any that was several degrees higher than 72°C would have been rewarded. **b** S/he states that the primers are single-stranded and that their base sequences are complementary to those on the **c** original strands. Although there is no mention of function, there is still enough information here to score 2 marks. **d** Like student A, this student has only explained part of the term in quotation marks and so scores only 1 mark. **e** Here, s/he really only explains again (and in some detail) what 'thermostable' means, and fails to score. There is no description of an *advantage*.

ⓔ **This is an example of a question that is 'easy if you know the answers'. There are no difficult ideas and no complex data. Did *you* know the answers? Student A scores 3 marks (grade D) and student B scores score 4 marks (grade C).**

Question 5 Gene cloning technology and its applications

Figure 1 shows how the action of enzyme **X** cuts a molecule of DNA.

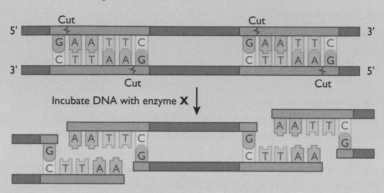

Figure 1

(a) (i)   Name the type of enzyme that cuts DNA in this way. (1 mark)

   (ii) What name is given to the places where the enzyme makes the cuts? (1 mark)

   (iii) Explain the importance of the type of cut made by this enzyme. (3 marks)

(b) Suggest why it may be preferable to obtain a gene from mRNA, rather than by cutting it from DNA. (2 marks)

Total: 7 marks

ⓔ Parts (a)(i) and (ii) each require a name; there is no need to write a sentence to answer either of them. Watch the mark tariffs in (a)(iii) and (b). You need to either give more than one concept or to give more than one explanation of any concept you introduce.

Student A

(a) (i)   Restriction **a** enzymes

   (ii) Sticky ends **b**

   (iii) It makes a jagged cut so that the two ends will fit together **c** again.

(b) This is because it will give a gene with no non-coding DNA **d**.

ⓔ **2/7 marks awarded a** This answer is acceptable, despite the term 'endonuclease' not being used. **b** Unfortunately, sticky ends are the product of digestion, not the name of the restriction site. **c** This answer does not gain credit as it shows no A-level understanding. If they 'fit together again', what was the point of cutting them apart in the first place? **d** This answer gains 1 mark but the student did not use the mark tariff intelligently. If s/he had explained *why* we would get no non-coding DNA, s/he would have gained the second mark.

AQA A2 Biology

Student B

(a) (i) Restriction endonuclease **a**

(ii) Restriction sites **a**

(iii) The two ends that are produced have complementary shapes and so fit together again. They also have complementary base sequences **b**.

(b) If the gene is active, there will a large amount of mRNA **c** in the cell. Also if you use mRNA, the introns have been removed **d** during post-transcriptional editing **d** but they would be in the DNA.

🅔 **5/7 marks awarded** **a** These two answers are correct. **b** The student has finally made one valid point but fails to provide a genuine explanation of the importance of this type of cut. **c** This answer is correct and gains a mark. **d** Each of these points is correct and would gain a mark, but this student cannot gain more than the 2 marks available for the question.

🅔 **Students often become confused over questions that, indirectly, ask for a comparison, such as part (b). In general, if an examiner can finish your sentences with either 'but not in the other' or 'but the other one does not', you will gain the mark. Both students gained marks in (b) for this reason. Neither student really explains the importance of jagged cuts in (a)(iii). Examiners were looking for concepts such as: they have unpaired bases; called sticky ends; which will bind to *other* complementary base sequences; so can be used to splice DNA together; e.g. inserting a gene into a plasmid. Student A scores 2 marks (grade U) and student B scores 5 marks (grade B).**

Question 6 The control of gene action

Small interfering RNA (siRNA) is a type of RNA that is important in 'silencing' genes.

(a) Give two differences between siRNA and:

    (i) **DNA**   (2 marks)

    (ii) **mRNA**   (2 marks)

(b) Huntington's disease is a disorder in which the protein produced by a mutant gene causes progressive death of cells in the brain. The cells of sufferers from this condition frequently contain one mutant gene and one normal gene. Figure 1 shows how siRNA could be used in the treatment of such conditions.

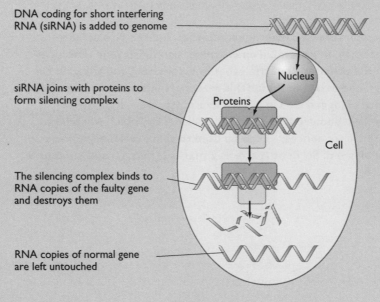

Figure 1

Use the diagram, and your knowledge of siRNA, to suggest how siRNA might one day be used to treat Huntington's disease.   (4 marks)

                                               Total: 8 marks

ⓔ Students often rush into questions without getting an overview of them. In a question like this, where the stimulus material occurs late in the question, it is a good idea to look at it to see if it contains anything that might help you to answer the earlier parts of the question. In this case, there is — see if you can spot it. You are given a lot of information in part (b) but you will not be rewarded for describing it. You are asked to apply the information to a specific context.

Student A

(a) (i) siRNA is smaller than DNA **a** and it is single-stranded **b**.

(ii) It is smaller than mRNA **c**.

(b) DNA is introduced into the nucleus that codes for siRNA **d**. The siRNA binds with proteins to form a silencing complex **d** that binds to RNA copies of the faulty gene **d** and destroys them **d**.

e **2/8 marks awarded** **a** This is correct, gaining 1 mark but this statement **b** is not correct. **c** This is also correct. **d** All these points simply repeat what is shown in the diagram. This student fails even to mention Huntington's disease, let alone explain how the treatment might work. S/he gains no marks for part (b).

Student B

(a) (i) DNA contains thymine but siRNA contains **a** uracil. siRNA is single-stranded **b**.

(ii) mRNA is a bigger molecule **a** that has hydrogen bonds **b**.

(b) DNA that codes for siRNA is introduced and siRNA is produced that is specific to the mRNA made by the Huntington gene **c**. When the siRNA joins with proteins to form a silencing complex, the complex joins to the Huntington mRNA **c** and destroys it **c**. This means that the Huntington mRNA cannot be translated into the Huntington protein **c** and so this would reduce the symptoms of this terrible disease.

e **6/8 marks awarded** **a** These statements are correct but **b** these are not. **c** Each of these statements is relevant, relates to Huntington's disease and scores a mark.

e **Unfortunately, in part (a), both students seem to believe that siRNA is single-stranded, despite it being shown as double-stranded in Figure 1. Did you spot this? Part (b) is more demanding. With such questions, make sure that you use the information supplied and do not merely restate it, as Student A did. Student A scores only 2 marks (grade U), while Student B scores 6 marks (grade B).**

Question 7 **Auxins**

The graph in Figure 1 shows the effect of different concentrations of an auxin on the growth of buds and shoots.

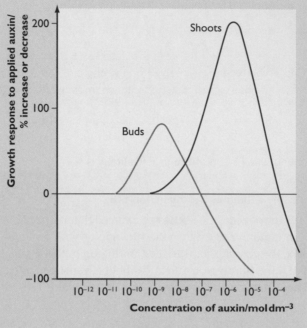

Figure 1

(a) **What is auxin?** (2 marks)

(b) **Describe what the graph shows about the effect of auxin on the growth of shoots.** (2 marks)

(c) **Use the information to suggest how a growing shoot inhibits the growth of side branches.** (3 marks)

Total: 7 marks

ⓔ You are unlikely to have seen this graph before. Don't panic in an examination; this type of question is testing your AO2 skills. Part (b) is the type of question you will commonly see in AS papers but rarely in A2 papers. Part (c) is a more demanding test of AO2 skills; you need to be careful with your wording in explaining it.

Student A

(a) Auxin is a plant hormone **a** that stimulates growth.

(b) Went showed **b** that increasing concentrations of auxin caused more growth, up to a maximum. Then there was no further increase in growth.

(c) Shoots grow in quite high concentrations of auxin, but buds are inhibited **c** by these concentrations.

(e) **2/7 marks awarded** **a** Use of the word hormone is strictly incorrect but an examiner would ignore this and award 1 mark for the concept of a chemical that stimulates plant growth. **b** In trying to show that s/he had learnt some experimental work, this student has failed to answer this question. It asked for a description of the graph. **c** The student has made a valid comparison, for 1 mark, but has failed to use any figures from the axes and failed to relate the answer to the question.

Student B

(a) Auxins are plant growth regulators **a**. They diffuse from the tips of roots and shoots and affect the elongation of cells **a** behind these tips.

(b) At a concentration of $10^{-9}\,mol\,dm^{-3}$ **b**, auxin starts to stimulate the growth of shoots. At a concentration of 10^{-6} the growth response reaches a maximum of 200% increase after which further increases in auxin concentration cause a decrease in growth until at 10^{-4} **c** it inhibits growth.

(c) Buds are stimulated to grow by lower concentration of auxins than shoots. Near the tip of a growing shoot the auxin concentration is 10^{-7} to $10^{-5}\,mol\,dm^{-3}$ **d**. The graph shows that this high auxin concentration inhibits the growth of buds **d**.

(e) **6/7 marks awarded** **a** Each of these statements is valid and scores 1 mark. **b** Having given the units once, examiners will accept values without units in the rest of this answer to part (b). This student has given a full description of the trend and **c** made specific reference to values where the gradient of the curve changes, scoring both marks. **d** Both of these points are valid, scoring 2 marks.

(e) **Parts (a) and (b) offer marks for fairly straightforward skills. Despite this, student A has scored badly, showing poor use of the mark tariff in (a) and answering a different question from the one set in (b). Part (c) is more demanding but student B made a very good attempt to use the information to answer the question. Student A scores 2 marks (grade U/E) and student B scores 6 marks (grade A).**

Question 8 Skeletal muscle

The sliding-filament theory is believed to offer the best explanation of muscle contraction.

(a) Figure I shows part of a relaxed myofibril.

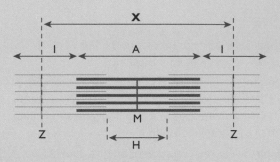

Figure 1

   (i) Name the region labelled **X**.   (1 mark)

   (ii) Re-draw the diagram to show the appearance of the myofibril following contraction.   (3 marks)

(b) Describe the roles of calcium ions and ATP in bringing about contraction of skeletal muscle.   (4 marks)

(c) Give two differences between slow-twitch and fast-twitch fibres.   (2 marks)

Total: 10 marks

ⓔ In (a)(ii), you need to measure the lengths of the lines you draw very carefully. Parts (b) and (c) are tests of AO1 with a high mark tariff, which would be rare in BIOL5 questions.

Student A

(a) (i) Muscle cell **a**

   (ii)

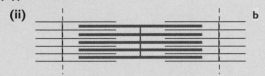

            b

(b) Calcium ions open the binding sites on the actin **c** and allow troponin **d** to bind. ATP supplies the energy **e** for the contraction.

(c) Slow-twitch fibres contract more slowly than fast-twitch fibres and with less force **f**.

ⓔ **4/10 marks awarded a** This student clearly has no idea that a myofibril is part of a muscle fibre (muscle cell), so part X cannot be a muscle cell. **b** The student has drawn a narrower sarcomere, which scores 1 mark. Although s/he has drawn the myosin filaments the same length as in the original, s/he has drawn the actin filaments shorter. As a result, s/he has an H zone that is the same width as the original, when it should be narrower. **c** S/he scores 1 mark for identifying that the binding sites are on the actin, but **d** incorrectly identifies troponin and **e** provides a superficial answer to the role of ATP. **f** S/he gives two valid functional differences in (c).

Student B

(a) (i) Sarcomere **a**

(ii)

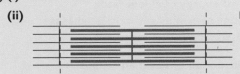

(b) Calcium ions allow tropomyosin **c** to bind with the actin. ATP releases energy to the myosin, which causes the myosin **d** to move and pull the actin filaments **e**.

(c) Slow-twitch fibres have more mitochondria **f** and more myoglobin **f** to supply oxygen.

ⓔ **8/10 marks awarded** **a** This answer is correct. **b** The lengths of the actin and myosin filaments are the same as in the original, and the sarcomere is narrower as is the H zone, scoring all 3 marks. **c** Calcium ions have the opposite effect to that which this student describes. **d** Although s/he fails to make clear that it is the myosin heads that are involved, she just makes the point that energy from ATP is used in some sort of movement of myosin and that **e** this pulls on actin filaments. **f** S/he gives two valid structural differences in (c).

ⓔ **This is a question in which both understanding and recall of detail are important. There is no shortcut: you must learn the structure and function of skeletal muscle. Student A scores 4 marks (grade E), while Student B scores 8 (grade A).**

Question 9 Receptors and transmission of information through the nervous system

Receptors transduce one form of energy into the electrochemical energy of a generator potential and an action potential. Nerve impulses may travel along neurones and across synapses to produce responses or they may be inhibited by action potentials from other neurones.

(a) Figure 1 shows the apparatus used in an investigation into the action of Pacinian corpuscles. Figure 2 summarises the results obtained.

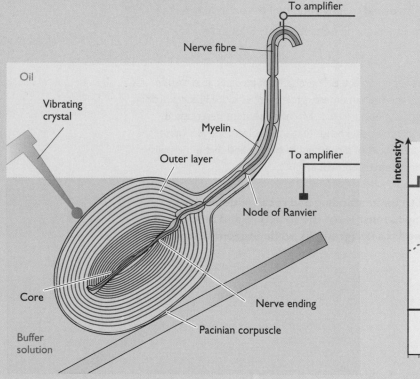

Figure 1

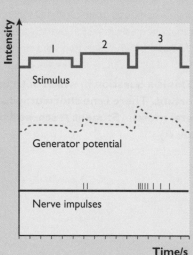

Figure 2

(i) How does the Pacinian corpuscle convert the vibrations of the crystal into a generator potential? (3 marks)

(ii) Suggest how the variations in the stimulus shown in Figure 2 could be generated using the above equipment. (2 marks)

(iii) Use your knowledge of the nature of nerve impulses to explain the results shown in Figure 2. (3 marks)

(b) Figure 3 and the table beneath it show how several sensory neurones can influence a single motor neurone.

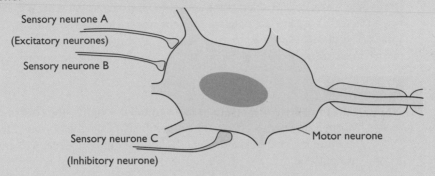

Figure 3

Neurone	Action potential in identified neurone				
Sensory neurone A (excitatory)	✗	✓	✗	✓	✓
Sensory neurone B (excitatory)	✗	✗	✓	✓	✓
Sensory neurone C (inhibitory)	✗	✗	✗	✗	✓
Motor neurone	✗	✗	✗	✓	✗

(i) Explain why a nerve impulse can only cross a synapse in one direction.   (3 marks)

(ii) Explain the results shown in the table.   (4 marks)

Total: 15 marks

ⓔ This is a fairly long question and there is a lot of unfamiliar stimulus material to interpret. You must practise AO2 and AO3 skills so that you are confident in dealing with questions like this under examination conditions. You might look for the sub-questions that test recall and answer those first. You might find part (b)(i) easier to answer than part (a)(iii) and choose to answer (b)(i) before (a)(iii). Analysing questions like this is part of the examination strategy you should be developing.

Student A

(a) (i) The vibrations of the crystal squash the Pacinian corpuscle **a** and the pressure causes the nerve ending to let ions **a** in and out **a**.

(ii) The crystal vibrates and the oil absorbs some of the energy **b** of the vibration. If you used less oil, **c** the vibrations would be bigger.

(iii) When the stimulus gets larger, the generator potential also gets bigger and you get more action potentials. This is because the action potential can't get bigger **d** — it's all-or-nothing.

(b) (i) The ends of the pre-synaptic neurones contain vesicles of neurotransmitter. When an action potential reaches the end, these vesicles release their neurotransmitter and it diffuses **e** across to the post-synaptic membrane. It can only pass one way because of the **f** concentration gradient.

(ii) When several neurones act independently like this on another neurone, the effect is called summation. You only get an action potential in the motor neurone when A and B act together **g**. If A and B act alone, or with C, there is no action potential **g** in the motor neurone.

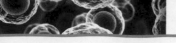

ⓔ 5/15 marks awarded a Although this student appears to have some understanding of how Pacinian corpuscles work, all these statements are too vague to score marks. **b** and **c** The student scores both marks for (a)(ii). Most of the answer to (a)(iii) is descriptive but **d** the student does explain why an increase in a generator potential does not produce an increase in the action potential, scoring 1 mark. The student explains **e** how neurotransmitters cross synapses and **f** why they only move in one direction, scoring 2 marks. **g** The points s/he makes in (b)(ii) are descriptions, not explanations.

Student B

(a) (i) The pressure of the crystal is transmitted through to the core of the Pacinian corpuscle where it affects the permeability of the membrane. Pressure-sensitive sodium ion channels are opened **a** and sodium ions enter **a** to create a generator potential **a**.

(ii) One could apply a larger voltage to make the crystal vibrate more **b**.

(iii) With more stimulation, the increased pressure opens more **c** sodium ion channels and the generator potential is increased. Because the generator potential generates the action potential, there ought to be a bigger action potential as well. But action potentials are 'all-or-nothing' **c** and so, instead of getting a bigger action potential, you get **c** more action potentials.

(b) (i) The synaptic knob of the pre-synaptic neurone contains vesicles that produce the neurotransmitter. When an action potential arrives at the synaptic knob, calcium ions enter and cause the vesicles to move to the surface and release their neurotransmitter into the synaptic cleft. It crosses the synaptic cleft and binds to protein receptor molecules on the post-synaptic neurone. This causes an action potential to be generated here and the nerve impulse is transmitted along the second neurone. **d**

(ii) Neurones A or B do not cause an action potential in the motor neurone on their own because they do not release enough neurotransmitter and so the threshold **e** is not reached. When both have an impulse and both release neurotransmitter, their combined effect is enough to pass the threshold **e** and start an action potential. When either A or B or both A and B have an impulse with C, the inhibitory effect of C is enough to counteract A and B **e** — together or separately. This is called summation.

ⓔ 10/15 marks awarded a Each of these points scores 1 mark and shows this student has learnt detail. **b** This valid suggestion scores 1 mark. **c** This student gives an excellent account in (a)(iii) and scores all 3 marks with these statements. S/he also gives an excellent account of how synaptic transmission takes place but **d** does not actually answer the question, which was about why transmission occurred in only one direction. The examiner could easily re-write what the student has written to score all 3 marks but is not permitted to do this. Unfortunately, this answer scores no marks. **e** In (b)(ii), the student has given three valid answers for 3 marks.

ⓔ The question requires a detailed understanding of Pacinian corpuscles and how transmission of nerve impulses takes place. Student A failed to provide this detail. Student B did, but failed to gain 3 marks in (b)(i) because s/he didn't adapt her/his knowledge to the question. Before rejoicing that you know the theory and writing at length, ask yourself, 'Am I answering the question the examiner has set me?'. Student A scores 5 marks (grade E) and Student B scores 10 (grade B).

Question 10 **Homeostasis**

Within set limits, humans are able to maintain a plasma glucose concentration and high body temperature that vary only slightly.

(a) (i) Explain the role of pancreatic hormones in maintaining the plasma glucose concentration within set limits. (6 marks)

(ii) Explain why these pancreatic hormones are able to target liver and skeletal muscle cells in particular. (4 marks)

(b) Explain the benefit of being able to maintain a constant, high body temperature. (5 marks)

Total: 15 marks

ⓔ You will not get a structured essay like this in BIOL5 – they appear in BIOL1 and BIOL4. Answering this question, however, will help you practise writing the 16-marks worth of content for the essay question in BIOL5.

Student A

(a) (i) When there is too much glucose in the blood, the pancreas **a** produces insulin to bring the level down **b**. When there isn't enough glucose in the blood, the pancreas produces glycogen **c** to raise the level. These two hormones maintain the level of glucose in the blood within set limits **d**.

(ii) Insulin can target liver and muscle cells because there are **e** receptors on these cells, but there aren't receptors **f** on other cells.

(b) Maintaining a constant, high body temperature means that you don't depend on the environment **g** to warm you up or cool you down. As the temperature **h** is always the same, all the processes in the body can carry on at the same rate because all biological processes **i** are affected by temperature.

ⓔ **2/15 marks awarded a** This student mentions the pancreas but, at A-level, an examiner would expect identification of the islets (of Langerhans). **b** S/he correctly identifies the action of insulin, gaining 1 mark, but provides no further information about how insulin exerts its effect. **c** S/he confuses glucagon with glycogen and **d** fails to add any further detail in the rest of the answer. Although in (a)(ii) s/he shows some understanding, s/he gains no marks. **e** S/he fails to mention either that receptors are protein/glycoprotein or that they are specific and **f** makes a statement later that fails to specify which receptors are absent. **g** In answering (b), the student realises that endothermy allows independence from the environment, gaining a mark. **h** S/he fails to make clear whether s/he is referring to environmental or body temperature and **i** makes a statement about biological processes that lacks any information about enzymes that you would expect of an A-level student.

Student B

(a) (i) If the plasma glucose concentration exceeds set limits, β-cells **a** in the islets of Langerhans **b** secrete insulin **a**. Insulin stimulates liver cells **c** to absorb glucose **d** and convert it to glycogen **d**. Glucagon **e**, secreted by α-cells **e**, has the opposite effect.

(ii) Receptors on the surface of liver and muscle cells have a specific shape **f** to which molecules of insulin and glucagon can bind. These receptors are not present on other cells **g**.

(b) A constant, high body temperature means that humans are independent of their environment **h** and can colonise inhospitable areas of the planet. A body temperature of 37°C means that the enzymes of our body are always working at their optimum **i** and so all the processes are carried out quickly. If our body temperature were higher than this, the enzymes would be **j** denatured.

ⓔ 10/15 marks awarded a In (a)(i), this student identifies correctly the hormone and the cells that produce it in **b** the islets. **c** S/he goes on to identify a target organ of insulin and **d** two of its effects on the target. **e** With little other detail, s/he correctly identifies the antagonist and its origin. S/he scores 5 marks in (a)(i). S/he scores 2 marks in (a)(ii) by **f** making clear that the shape of the receptors identifies a target cell and that **g** these receptors are not present on other cells. In (b) s/he scores 3 marks by **h** providing one benefit of endothermy, **i** identifying that the effect of temperature on enzymes is critical and **j** providing further explanation.

ⓔ Most students think they understand homeostasis. Often they do understand the main principles but, as shown in one example above, frequently miss marks by providing answers that fail to show the learning that examiners expect of an A-level student. Student A scores just 2 out of 15 marks (grade U) and Student B scores 10 (grade A/B).

Knowledge check answers

1 A change in an organism's internal or external environment.

2 A taxis involves *movement* towards or away from a stimulus. A tropism involves *growth* towards or away from a stimulus.

3 An action potential occurs in a nerve cell whereas a generator potential occurs in a receptor cell.

4 Each transduces a different form of energy.

5 Blood pressure increases because contraction of skeletal muscles forces more blood to the heart. pCO_2 increases because increased respiration by active muscles produces more CO_2.

6 In the wall of the right atrium.

7 Since they move down a concentration gradient through ion channels, it must be facilitated diffusion.

8 The ability to distinguish objects that are close together — the amount of detail you perceive.

9 (a) Several rods synapse with a single bipolar cell. (b) Several receptors stimulate a nerve cell, ensuring that the threshold stimulation is reached.

10 They are absent at the fovea, rise in number to $160000/170000\,mm^{-2}$ at 10° to 20° and then decrease in number to the edge of the retina.

11 If you look directly at the star, its image falls on cones which are insensitive to dim light. If you look to one side, its image falls on rods; summation ensures the threshold is reached in the bipolar cell.

12 They bring about the opposite effect.

13 Against their concentration gradient (from lower to higher concentration).

14 Figure 16 shows that a new action potential cannot be initiated until about 2 ms after the start of the last one, thus limiting the frequency of impulses.

15 Repolarisation involves active transport, which uses energy from the hydrolysis of ATP; this is an enzyme-controlled reaction and temperature affects enzyme activity.

16 Resistance is less if the diameter is greater (think of resistance to blood flow in arteries and capillaries).

17 Myelinated; depolarisation jumps from node to node.

18 A motor neurone.

19 Energy, from the hydrolysis of ATP, is used in making neurotransmitter; mitochondria make ATP.

20 The post-synaptic membrane has protein receptors that are complementary to the shape of only one of acetylcholine, GABA or noradrenaline.

21 (a) No; threshold stimulation is not reached. (b) Yes; summation results in the threshold being exceeded. (c) No; the opposite effects of the excitatory (A) and inhibitory (C) neurones result in little change to the membrane potential of D.

22 It saves energy if the cell recombines the partly hydrolysed components rather than starting from scratch.

23 Only the pre-synaptic neurone has vesicles containing neurotransmitter; only the post-synaptic membrane has receptors for the neurotransmitter; the concentration gradient of neurotransmitter is from pre- to post-synaptic membrane.

24 A simple reflex is inborn/unlearned and is one in which the stimulus always produces the same response.

25 Relay neurones in the spinal cord pass impulses to neurones in the brain.

26 The strength of response would reach a peak earlier and would be shorter lived in the curve representing nervous control.

27 They are lipid-soluble and the membrane is largely made of phospholipids.

28 The coleoptile grows towards the light if the tip is intact and light can get to it. Cutting off the tip or stopping light getting to it results in no growth towards the light.

29 Mica stops passage of chemicals and prevents the growth towards light. Gelatin allows passage of chemicals and growth towards light occurs normally.

30 No curvature occurred if the tip was replaced centrally but if placed on one side curvature occurred away from that side.

31 Went's experiment; he used agar blocks containing the chemical, identified the chemical as auxin (IAA) and had a control (no auxin).

32 Auxin (IAA) accumulates on the lower side of both root and shoot. In the root, a high concentration inhibits cell elongation so the root curves downwards. In the shoot, a high concentration stimulates cell elongation so the shoot curves upwards.

33 The full force of contraction is transmitted to the bone; none is lost in stretching the tendon.

34 It contains different tissues, e.g. muscle tissue, connective tissue and nervous tissue, as well as organs such as blood vessels.

35 The H zone contains myosin filaments, which are thicker than the actin filaments in the I band. The rest of the A band contains both actin and myosin, so appears even darker than the H zone.

36 (a) No, they have all stayed the same length. (b) The sarcomere, I band and H zone are narrower because the actin filaments have been pulled into the centre of the sarcomere.

37 Calcium ions cause neurotransmitter to be released from the motor neurone and cause tropomyosin to move, exposing the myosin binding sites on the actin.

38 (a) From 'stored' ATP and then from creatine phosphate, which last the length of the exercise. The athlete uses aerobic respiration after the race has finished to produce more ATP. (b) From 'stored' ATP, then creatine phosphate, then aerobic respiration. More ATP is produced by aerobic than anaerobic respiration, enabling the athlete to complete the marathon.

39 (a) Fast twitch — this event is short-lived and requires strong contractions. (b) Slow-twitch — this event takes a long time and requires energy release over a sustained time.

40 (a) Fast-twitch. (b) Slow-twitch.

41 If its water potential became more negative, cells might lose water to the tissue fluid by osmosis. If its water potential became less negative, cells would gain water from the tissue fluid and might burst.

42 Glucose is a respiratory substrate (some organs, like the brain, cannot store it). Glucose lowers the water potential of the blood.

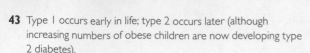

43 Type 1 occurs early in life; type 2 occurs later (although increasing numbers of obese children are now developing type 2 diabetes).

44 Low levels of oestrogen allow FSH levels to rise; rising levels of oestrogen inhibit FSH secretion (negative feedback); even higher levels of oestrogen stimulate FSH secretion (positive feedback). (b) High levels of oestrogen stimulate LH production (positive feedback).

45 (a) CUA. (b) Asp (aspartic acid).

46 It might change the tertiary structure of the protein so that it no longer carries out its function effectively.

47 (a) As it grows, it puts pressure on surrounding tissues and organs. (b) Cells from a malignant tumour (cancer) break away and grow in other parts of the body (metastasis). This does not happen with a benign tumour.

48 In the direction of transcription, they are located in front of the gene they regulate.

49 Multipotent — they will produce blood cells only.

50 They have a shape that is complementary to part of the RNA polymerase molecule.

51 (a) Yes, mRNA is produced from it. (b) No, the mRNA is destroyed before it is translated.

52 (a) A short length of single-stranded DNA (ssDNA) with a known base sequence that is tagged with a radioactive or fluorescent marker. (b) It hybridises with another ssDNA that has a complementary base sequence and is commonly used to locate a gene.

53 The restriction site has a specific base sequence that is complementary to the active site of the enzyme (think back to the lock-and-key or induced-fit models of enzyme action).

54 The phosphate groups (PO_4^{2-}).

55 (a) Five times. (b) Six times. (Try both with a piece of string and a pair of scissors.)

56 The mRNA has had the introns removed; the gene still has the introns intact.

57 The unpaired bases will pair with DNA with a complementary base sequence, allowing two pieces of DNA to be joined, e.g. a gene and a plasmid.

58 If DNA from a technician or from a foreign body has a complementary base sequence to that of the primer, it will be amplified by the PCR as well.

59 It enables DNA polymerase to begin replicating the DNA templates.

60 Recombinant refers to DNA produced by splicing DNA from two different sources. Transgenic refers to one organism with some DNA from another organism.

61 There were nine colonies (top of picture) of which three were labelled (bottom of picture), so 33.3% were transformed.

62 The sequence of the fragments in Figure 65 is ATGCCTGCA. Since the fragments are complementary to the original DNA, the base sequence of the original DNA was TACGGACGT.

63 Because, unlike coding DNA, non-coding DNA shows great variation from person to person.

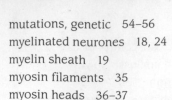